INTRODUCTION TO WELL LOGS & SUBSURFACE MAPS

Jonathan Evenick

Disclaimer

The recommendations, advice, descriptions, and the methods in this book are presented solely for educational purposes. The author and publisher assume no liability whatsoever for any loss or damage that results from the use of any of the material in this book. Use of the material in this book is solely at the risk of the user.

Copyright © 2008 by
PennWell Corporation
1421 South Sheridan Road
Tulsa, Oklahoma 74112-6600 USA

800.752.9764
+1.918.831.9421
sales@pennwell.com
www.pennwellbooks.com
www.pennwell.com

Marketing Manager: Julie Simmons
National Account Executive: Barbara McGee

Director: Mary McGee
Managing Editor: Stephen Hill
Production Manager: Sheila Brock
Production Editor: Tony Quinn
Cover Designer: Lori Duncan
Book Layout: Lori Duncan

Library of Congress Cataloging-in-Publication Data

Evenick, Jonathan.
 Introduction to well logs and subsurface maps / Jonathan Evenick.
    p. cm.
 Includes bibliographical references and index.
 ISBN 978-1-59370-138-3
1. Geophysical well logging--Data processing. 2. Geological mapping--Data processing.
I. Title.
TN871.35.E94 2008
622'.1828--dc22

                                                              2007041130

Printed in the United States of America

5 6 7 8   16 15

INTRODUCTION TO WELL LOGS
& SUBSURFACE MAPS

CONTENTS

List of Figures. ix

Acknowledgments. xv

Preface xvii

1. Introduction to Well Logs and Terminology. 1

Well Information. 2

API Numbering. 5

   State code 5

   County code 6

   Well number 6

   Sidetrack code 7

   Event code. 7

Subsurface Thicknesses 8

Exercises. 12

Questions 15

2. Basic Well Logs and Log Signatures 17

Basic Well Log Types 17

   Gamma Ray (GR) 18

   Resistivity 20

   Spontaneous Potential (SP) 20

   Photoelectric (PE) 20

   Neutron. 21

   Density 21

   Dipmeter 22

   Caliper. 22

   Sonic (Acoustic) 22

   Temperature. 23

   Log Signature Patterns 23

Exercise 25

Questions 31

3. Introduction to Subsurface Maps and Contouring 33

Contouring 34
Exercises 37
Questions 40

4. Structural and Stratigraphic Interpretations 41

Picking Methodology 41
Structural Interpretations 43
Stratigraphic Interpretations 45
Exercises 47
Questions 55

5. Structure Contour Maps 57

Exercises 60
Questions 72

6. Thickness Maps 73

Exercises 78
Questions 90

7. Facies Maps 91

Exercises 95
Questions 106

8. Trend Surface Maps 107

Three-Point Problems 119
Exercises 112
Questions 117

9. Trend Surface Residual Anomaly Maps 119

Making a TSRA Map 121
Exercises 124
Questions 136

10. Hydrologic Maps and Injection Wells 137

Hydrologic Maps. 137
Injection Wells 140
Exercises. 143
Questions 148

11. Formation Fluid Interpretation and Hydrocarbon Reservers 149

Water Saturation Calculations. 152
Calculating Hydrocarbon Reserves 154
Exercises. 158
Questions 160

12. Mining Maps 161

Calculating Mining Resources and Reserves 163
Exercise 165
Questions 167

13. Cross Sections. 169

Cross Section Types 169
    Contour slicing. 172
    Vertical exaggeration 174
    Balanced cross sections 175
Exercises. 178
Questions 184

Appendix: Swan Creek Term Project 185

Exercise 188
Questions 209

Glossary. 211

List of Abbreviations. 221

Bibliography 225

Index 227

LIST OF FIGURES

Figure 1–1 A well header or typical first page of a well log......................3

Figure 1–2 Common linear and logarithmic grids and track numbers............4

Figure 1–3 Two wells with the same name, but drilled in different states..........6

Figure 1–4 Two wells with two sidetracks off the main borehole.................7

Figure 1–5 The well on the left was deepened and one of the sidetracks
was plugged back; whereas the well's main borehole on the
right was plugged back..8

Figure 1–6 An illustration of three wells, one vertical (A) and two deviated
(B and C), that penetrate a dipping unit with a
constant thickness..9

Figure 1–7 An example of how misusing thickness values in volume
estimates can be problematic...................................11

Figure 1–8 Fill in the following values for the locations noted in the three logs 12

Figure 1–9 Fill in the API numbers for the following wells......................13

Figure 1–10 Calculate the TVT, TST, and TVDT in the following example........14

Figure 2–1 Chart of common logs and uses.................................18

Figure 2–2 Generalized well log signature patterns 19

Figure 2–3 Generalized well log signature patterns 24

Figures 2–4 through 2–9 Identify the lithologies in the following well logs 25

Figure 3–1 A temperature map for the continental U.S. 35

Figure 3–2 Structure contour map with a thrust fault (fault break)...............35

Figure 3–3 Two-dimensional views of differently spaced contours...............36

Figure 3–4 Contour the following data and complete the two profiles 37

Figures 3–5a–c Draw three structure contour maps to illustrate what the
décollement-related anticlines in the cross section may look like in
map view..38

Figure 3–6 Identify the lithologies in the well log 39

Figure 4–1 Cross sections through a structural high 42

Figure 4–2 Type log with multiple picks from Savannah, Georgia 43

Figure 4–3 Structure contour map showing multiple folds 44

Figure 4–4 A composite-type well log derived from two
             nearby wells that are faulted 45

Figure 4–5 An isopach map with an irregular pattern except for a minor NNE
             trending feature near the middle of the map that could be related to a
             structural feature. 46

Figures 4–6 through 4–13 Identify the lithologies in the following wells and correlate
             the wells on the provided cross section 47

Figure 5–1 Three-dimensional, wireline mesh representation of a structure contour
             map having an east-west trend in the northern part of the map and a
             northeastern trend in the southern portion of the map 58

Figure 5–2 Three-stacked structure contour maps illustrating an angular
             unconformity 59

Figures 5–3 through 5–15 Identify the lithologies in the following wells and
             calculate the depth to the top of the salt and bentonite units. 60

Figure 6–1 A structure contour, isochore, and isopach map of a simple fold 74

Figure 6–2 Cross section of a salt dome trap 75

Figure 6–3 A well log showing the net pay within a limestone and sandstone
             reservoir using a 4% cutoff neutron porosity 76

Figure 6–4 A common porosity conversion chart 77

Figures 6–5 through 6–17 Identify the lithologies in the following wells
             and calculate the depth to the top of the coal (Pick 1) and
             limestone (Pick 2) units 78

Figure 7–1 Transgressional (top) and regressional (bottom) environments
             indicating changes in sea level, subsidence or uplift rates,
             or rates of seafloor spreading 92

Figure 7–2 Inferred porosity, permeability, and depositional environments
             of common sedimentary rock types 94

Figure 7–3 Various types of stratigraphic traps: A) isolated; B) onlap
             unconformity; C) reef; D) up-dip pinchout; and E) unconformity 95

Figures 7–4 through 7–15 Identify the lithologies in the following wells
and calculate the depth to the top of the bentonite bed 95

Figure 8–1 Schematic 3-D view of two trend surface maps separated by
an angular unconformity 108

Figure 8–2 Solving a three-point problem and creating a trend surface map 110

Figures 8–3 through 8–7 Identify the lithologies in the following exploration
wells and calculate the depth to the top of the two salt horizons
and a bentonite bed 112

Figure 9–1 Stacked structure contour maps illustrating a general dip
to the lower left 120

Figure 9–2 Stacked trend surface residual anomaly maps that correspond with
structure contour surfaces 120

Figure 9–3 A first-order trend surface map and a corresponding trend surface
residual anomaly map generated by subtracting the structure
contour data (adjusted pick depth) from the trend surface 122

Figure 9–4 A general TSRA map flowchart 122

Figures 9–5 through 9–16 Identify the lithologies in the following wells
and calculate the depth to the top of an anhydrite interval 124

Figure 10–1 An idealized, uncased well log for saturated and
unsaturated strata. 138

Figure 10–2 A map of the water table and a cross section illustrating how
groundwater travels in the subsurface with respect to
hydraulic head 139

Figure 10–3 Schematic map and cross section views of an injection well
surrounded by producing wells 140

Figure 10–4 A 2-D view of the swept or sweep area after four days 142

Figure 10–5 A 2-D view of the swept or sweep area after 1,500 days 142

Figure 10–6 Contour the hydraulic head data and draw flow lines 143

Figure 10–7 This is a complex channel system that is a producing
hydrocarbon horizon. 144

Figure 10–8 Using the dominant lithologies from exercise 10–2,
shade in the area swept after 10 days 145

Figure 10–9 Using the dominant lithologies from exercise 10–2,
shade in the area swept after 100 days 146

Figure 10–10 Using the dominant lithologies from exercise 10–2,
              shade in the area swept after 500 days 147

Figure 11–1 Common geophysical responses to different types
              of formation fluids. 150

Figure 11–2 A structure contour map on top of a 120-m (394-ft) thick,
              hydrocarbon-filled interval 151

Figure 11–3 Two graphs of modeled production data (four years observed
              and 11 years estimated) of a well that has an economic limit
              of 6 bbl 156

Figure 11–4 Identify the lithologies in the well. 158

Figure 11–5 Draw the decline curve for a theoretical well using the
              production data, and calculate the economic reserves 159

Figure 12–1 A mineral resource system (source, reservoir, and seal)
              for a fault-controlled ore deposit. 162

Figure 12–2 Test wells indicated that the extensively fractured strata is
              flat-lying and that the ore is located in a 10-m thick section
              in the upper part of a carbonate unit. 165

Figure 12–3 Identify the lithologies in the well and contour the ore
              grades using a 3% interval. 166

Figure 13–1 Chart of cross section categories. Cross sections are subdivided
              by the amount of well and data control 170

Figure 13–2 Different cross section types. A marker unit (gray) generally
              thickens to the east and north 171

Figure 13–3 A detailed cross section along a given profile .173

Figure 13–4 An example of a cross section (same as fig. 13–3d) with
              different vertical exaggerations. 175

Figure 13–5 Balanced and unbalanced cross sections. 176

Figure 13–6 Standardize the given offshore well data to sea level (SL) and calculate
              the pick elevations. 178

Figures 13–7 through 13–10 Create four structure contour maps using the
              standardized data from exercise 13–1. 180

Figure 13–11 Create four profiles (X-X') and stack them in the provided
              cross section. 184

Figure A–1 Well location map for the Swan Creek field in
           northeastern Tennessee 186

Figure A–2 A type log through the reservoir rocks in the
           Swan Creek field with five picks.............................187

Figure A–3 Topographic and structure contour maps for Picks 1 and 2 188

Figure A–4 Structure contour maps for picks 1 through 3 189

Figure A–5 A-A' and B-B' cross sections 190

Figure A–6 Well log for 41-067-20011 191

Figure A–7 Well log for 41-067-20017 192

Figure A–8 Well log for 41-067-20022.....................................193

Figure A–9 Well log for 41-067-20023.....................................194

Figure A–10 Well log for 41-067-20029 195

Figure A–11 Well log for 41-067-20032 196

Figure A–12 Well log for 41-067-20036.....................................197

Figure A–13 Well log for 41-067-20037 198

Figure A–14 Well log for 41-067-20038 199

Figure A–15 Well log for 41-067-20039 200

Figure A–16 Well log for 41-067-20040.....................................201

Figure A–17 Well log for 41-067-20041 202

Figure A–18 Well log for 41-067-20043.................................... 203

Figure A–19 Well log for 41-067-20044..................................... 204

Figure A–20 Well log for 41-067-20045.....................................205

Figure A–21 Well log for 41-067-20046.................................... 206

Figure A–22 Well log for 41-067-20052.................................... 207

Figure A–23 Well log for 41-067-20054.................................... 208

ACKNOWLEDGMENTS

I would like to thank Tengasco for donating the digital well data used in the Swan Creek project and access to the field. I would also like to thank Nancy Meadows for editing help, and Robert D. Hatcher, Jr. (University of Tennessee), Gary Bible (Miller Petroleum), Craig Calvert (ExxonMobil), Mike Farrell (ExxonMobil), Omar Varela (ExxonMobil), Timo von Rudloff (Talisman Energy), and Holly Harrison (BP America Inc.) for various discussions regarding topics used in this workbook.

PREFACE

This book introduces different types of geophysical logs and subsurface maps that can be generated from basic well data, and subsurface problems that can be solved using geophysical logs and subsurface maps. "Hands-on" exercises will reveal how each map type is generated and what applications they may have. The exercises at the end of each chapter will introduce different types of wells and lithologies; please refer to chapter 2 until you become more comfortable with each log type and what it measures.

Subsurface mapping is a way to visualize various geologic and hydrologic features in any dimension from a 1-D cross section to a 4-D production map. All subsurface map types can be useful, but the key is to know what you are investigating and what map types are most appropriate. For example, to evaluate facies changes, a 2-D structure contour map may not be as helpful as a simple cross section. Most subsurface maps and data are digitally constructed, manipulated, and interpreted by geologists using advanced computer contouring algorithms and software packages. Given the size of datasets and inherent complexities in subsurface mapping, it is necessary to use computers to analyze and visualize data. It is, however, fundamentally important to understand what different maps represent in the subsurface and in three dimensions. Computers are capable of generating excellent maps and cross sections, but an experienced geologist with solid background can produce superior hand-drawn maps. Subsurface mapping and interpreting geophysical logs are non-unique arts. There can be several plausible solutions, but none of them have to be the correct answer. Keep in mind that a map is never complete and new data will always alter it and possibly an inferred geological paradigm. Good luck with the exercises and keep an open mind.

For the answer key, please contact the author at jevenick@hotmail.com.

1

INTRODUCTION TO WELL LOGS AND TERMINOLOGY

After a well is drilled, a string of geophysical instruments (called a *sonde*) is placed into the *borehole* to record the geophysical properties of the subsurface strata. The data are digitally recorded in a nearby logging facility (commonly a logging vehicle for land-based *rigs*) and transferred to a paper log and/or saved as a digital .las data file. It is very common to get both a digital and paper copy of the well. Paper logs give interpreters the extra advantage of being able to quickly view and shift logs to make quick correlations. Digital files, however, are more useful in deriving specific values and can be quickly rescaled and electronically transferred. Logging can also take place while drilling. This requires the geophysical logging instruments to be operated behind the drill bit, which can be highly advantageous in horizontal or highly deviated wells because it is difficult to properly log these wells after drilling is complete. Logging while drilling also enables almost instantaneous data transfer between the bottom of the borehole and the drillers or interpreters on the rig or in the office.

WELL INFORMATION

Well data are the control points from which regional stratigraphy can be derived and local or regional deviations can be identified that could correlate to structural or stratigraphic changes. Well data are ideally integrated into seismic reflection data, but in some areas no other datasets are available. Well data interpretation is a crucial step to understanding any subsurface problem. Well data can be extracted from the *well ticket* or the *well log.*

A well ticket is a synopsis of the important drilling information. It typically contains the well name, *API number,* location, driller, drilling dates, elevation, and stratigraphic correlations, or *picks.*

A well log is a detailed record of the geophysical and physical properties in and around the borehole obtained during drilling. The first page of a well log is called the well header (fig. 1–1), and all the basic drilling and logging information related to the well is listed there. The essential well information needed to start mapping are the well name, location, logging elevation, and stratigraphic correlations. Stratigraphic correlations are derived by interpreting the geophysical data and do not appear in the well header, but may be placed in the well ticket. Most logs are not run from ground level (G.L.), but rather from the *Kelly bushing* (K.B.) or *drilling floor* (D.F.). It is very important to read from which datum the log was measured.

The body of a well log is commonly divided into four *tracks* (fig. 1–2) and the geophysical data are plotted in these tracks. Track 1 is on the left side of the log and has the depth scale located to the right of it. This track has a linear scale and is used for *gamma ray, caliper,* and *spontaneous potential* logs. Tracks 2 and 3 are to the right of the depth column and can be either linear or logarithmic scales. Track 2 is to the immediate right of the depth column and track 3 is on the far right side of the log. Track 4 extends the length of tracks 2 and 3, and is primarily used for *resistivity* or *porosity.* Logs typically run in the second, third, and fourth tracks are *density, resistivity, sonic, neutron, temperature,* and *photoelectric factor.* The target type and regional geology will determine what type of logs or information is needed to solve or address the subsurface problem.

Company Name

Log Type

Filing No.		
Permit # **and** **API #**	Company ____ Company name ____ Well ____ Well name ____ Field ____ Field name ____ County ____ County name ____ State ____ State name ____	

Location:		Other Services:
Location information SEC._____ TWP._____ RGE._____		Other logs run

Permanent Datum: ____ Log elevation ____ Elev.: ____		Elev.: K.B. _____
Log Measured From _____, Ft. Above Perm. Datum Drilling Measured From ____ information ____		D.F. ____ G.L. Elevations

Date			
Run No.			
Type Log			
Depth-Driller			
Depth-Logger			
Bottom logged interval			
Top logged interval			
Type fluid in hole	Logging	Logging	Logging
Salinity, PPM Cl.	information	information	information
Density			
Level			
Max rec. temp., deg F.			
Operating rig time			
Recorded by			
Witnessed by			

Run	Bore-Hole Record			Casing Record			
No.	Bit	From	To	Size	Wgt.	From	To
		Borehole information				Casing information	

Fig. 1–1. A well header or typical first page of a well log. Lists logging and basic drilling information.

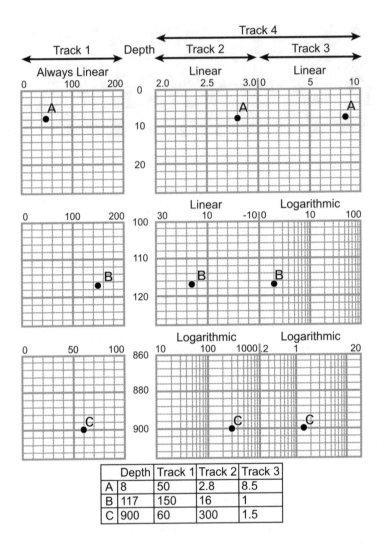

Fig. 1–2. Common linear and logarithmic grids and track numbers. Letters refer to the location of the readings. A grid for track 4 is not shown, but it can be either linear or logarithmic.

API NUMBERING

Every U.S. state has its own method for cataloging and sorting well data. Some states, such as Tennessee, use the permit number as the main well ID number, whereas Kentucky uses a well's record number. It is very common to refer to wells by name (i.e., J.Q. Public #1), but this is not a good practice if you are interested in collecting data from areas as small as a field. This is because well names can be very similar or even duplicated in different states or counties (fig. 1–3). Therefore, the American Petroleum Institute (API) set up a methodology for giving every well in the United States a unique ID number, or API number. An API number is made up of five sets of numbers (14 numbers) that correspond to a well's attributes (state drilled, county, well number, sidetrack, and event sequence). The API number is commonly listed as the first 10 numbers because a vertical well will have zeros for the last four numbers. *Deviated wells* that start or are *kicked* off a single borehole are typical in offshore drilling, making it necessary to list the last four numbers. An example API number for a vertical well is 31-007-14059-00-00.

State code

The first two digits of the API number represent the state where the well was drilled. If the bottom of a deviated well is in another state or county, only the state and county where drilling started, or *spudded*, is listed. The state codes correspond to the alphabetic listing of the 48 contiguous states plus the District of Columbia (08). Alaska (50) and Hawaii (51) are the only two states that do not follow this convention.

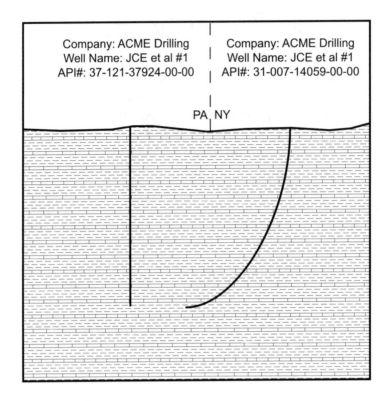

Fig. 1–3. Two wells with the same name, but drilled in different states. Note that one well terminates in PA, but has an API number indicating it was started in NY.

County code

The third through fifth digits of the API number represent the county where the well was drilled. County codes are commonly odd-numbered, but some even-numbered county codes do exist (e.g. Arizona, Wisconsin, Virginia, and New Mexico). This was designed for counties having more than 100,000 wells and to accommodate the formation of new counties. County codes are also used for offshore areas for both state and federal waters.

Well number

The sixth through tenth digits of the API number represent the unique and sequential number of wells drilled in the county. In some states, the well number is based on the permit number or record number. Wells drilled before API numbers were used are typically given numbers between 00000 and 10000. If a well has been reentered it may have an R placed after the well number with the corresponding number of times it has been reentered.

Sidetrack code

The eleventh and twelfth digits of the API number represent the *sidetrack* code. The sidetrack number typically corresponds to the number of sidetracks in the well (fig. 1–4). Sidetracks off other sidetracks are given numbers as well as sidetracks off the *master borehole*.

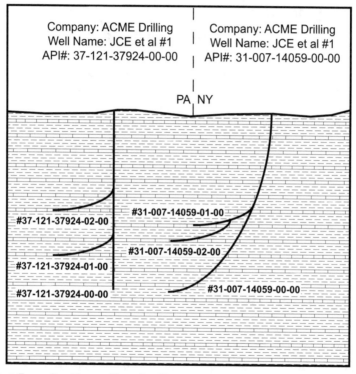

Fig. 1–4. Two wells with two sidetracks off the main borehole. Note that the sidetracks are numbered in the order they were drilled.

Event code

The thirteenth and fourteenth digits of the API number represent alterations to an individual sidetrack (fig. 1–5). The number corresponds to the order in which a well is deepened or partially plugged. This code has not been standardized and some companies only report R1.

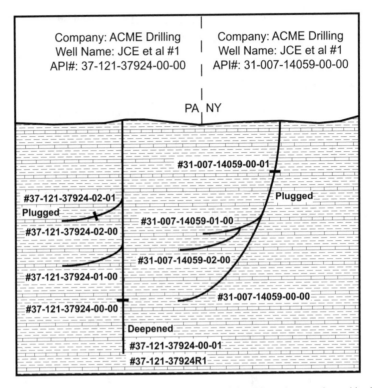

Fig. 1–5. The well on the left was deepened and one of the sidetracks was plugged back, whereas the well's main borehole on the right was plugged back.

SUBSURFACE THICKNESSES

In order to speak the same logging language, there are a few terms to describe exactly what thicknesses are being measured in the subsurface. The calculated thickness of an interval measured in a well log is called the *measured log thickness* (MLT; fig. 1–6). In flat-lying strata, this is equal to the *true stratigraphic thickness* (TST), but in dipping strata the MLT is larger than the TST (unless the well is deviated and perpendicular to the unit). *True vertical thickness* (TVT) is the vertical thickness of an interval and the *true vertical depth thickness* (TVDT) is the vertical thickness between the borehole entry and where it exited the interval. In a vertical well the MLT, TST, TVT, and TVDT are all equal, but they are not equal if the wellbore is deviated or if the strata are dipping. In the latter cases, *Setchell's equation* (1.1) can be used to calculate the TVT regardless of the borehole direction or deviation, and dip of the strata. The basic variables needed are the

MLT, the dip angle and azimuth direction of the borehole, and the dip angle and direction of the unit. The TST can be calculated from the TVT using equation (1.2) and the TVDT, if needed, can be calculated using equation (1.3).

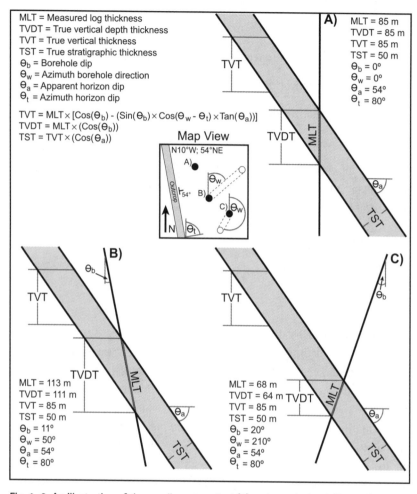

Fig. 1–6. An illustration of three wells, one vertical (A) and two deviated (B and C), that penetrate a dipping unit with a constant thickness. See the map view in the middle for well locations. Each well has the same TST, but different MLTs, TVDTs, and TVTs.

$$\text{TVT} = \text{MLT} \times [\text{Cos}(\theta_b) - (\text{Sin}(\theta_b) \times \text{Cos}(\theta_w - \theta_t) \times \text{Tan}(\theta_a))] \qquad (1.1)$$

where:

MLT = measured log thickness (calculated from log picks)

TVDT = true vertical depth thickness

TVT = true vertical thickness

TST = true stratigraphic thickness

θ_b = borehole dip (measured from vertical)*

θ_w = azimuth borehole direction (measured from north)*

θ_a = apparent horizon dip (measured from horizontal)*

θ_t = azimuth horizon dip (measured from north)*

*Borehole information can be found from the directional well survey and the stratigraphic information can be found from outcrop measurement, *dipmeter log*, *three-point problems*, or *structure contour* maps.

$$TST = TVT \times Cos(\theta_a) \qquad\qquad (1.2)$$

$$TVDT = MLT \times Cos(\theta_b) \qquad\qquad (1.3)$$

These terms and values are important when subsurface correlations and estimates are made in complex or deformed areas. Using the wrong thickness can mean the difference of billions of dollars. For example, if a 25-m (82-ft) thick horizontal beam were placed in the earth (fig. 1–7) with a cubic volume of 250,000 m³ (326,988 yd³) that represented an economic commodity with a market cost of $40/m³, the value of the beam would be $10 million. Now if this beam were placed horizontally at an angle of 30°, the TVT would equal approximately 29 m (95 ft) instead of 25 m (82 ft). If the same other dimensions were fixed because of property boundaries, using the TVT thickness would yield an inaccurate volume estimate of 290,000 m³ (379,306 yd³) and value of $11.6 million ($1.6 million more true value). Altering the angle to 45°, the volume would be increased to approximately 350,000 m³ (457,783 yd³) and the value would be $14 million. Hence, it is important when dealing with subsurface estimates to use the correct thicknesses. Note that the TVT is more commonly used in subsurface estimates than the TST because it is faster to multiply the TVT times the apparent width and length of the commodity than multiply the TST times the real width and length.

In terms of stratigraphic correlations, units dipping under 10° will not have a significant impact on the difference between the TVT and TST. In wells deviated more than 10° or units dipping greater than 10°, thicknesses should be corrected before any subsurface correlation work is attempted. Fortunately, most deviated wells have two depth scales: a log depth and a corrected true vertical depth (TVD). The TVD eliminates the need to manually correct the TVT.

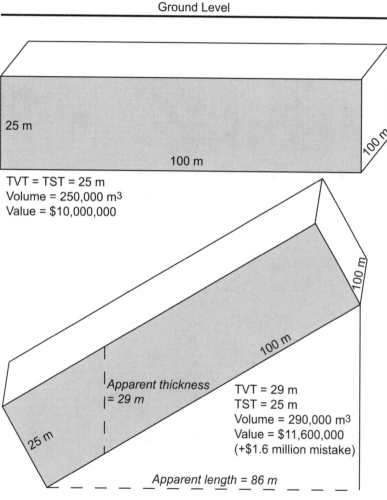

Fig. 1–7. An example of how misusing thickness values in volume estimates can be problematic. In this example, the TVT was incorrectly used with the length of the beam instead of using the TST. The TVT value can only be correctly used with the apparent length to yield an accurate volume estimate. In practice, this calculation method is typically easier to solve because it does not involve calculating the true length and width of the target.

EXERCISES

1–1) Fill in the following values for the locations noted in the three logs (fig. 1–8).

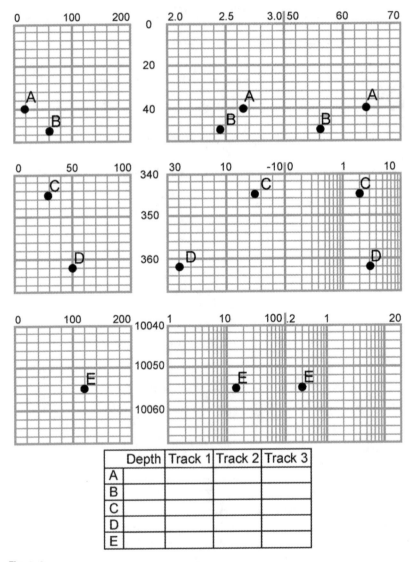

	Depth	Track 1	Track 2	Track 3
A				
B				
C				
D				
E				

Fig. 1–8

1–2)   Fill in the API numbers for the following wells. Note there is more than
one correct answer (fig. 1–9).

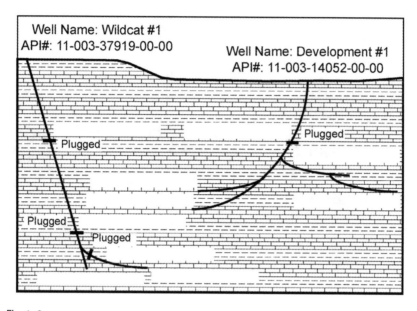

Fig. 1–9

1–3) Calculate the TVT, TST, and TVDT in the following example (fig. 1–10).

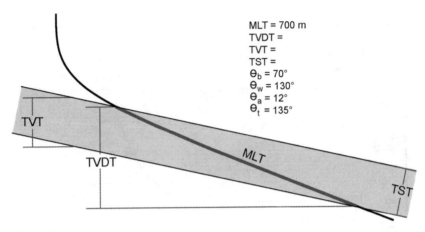

MLT = 700 m
TVDT =
TVT =
TST =
Θ_b = 70°
Θ_w = 130°
Θ_a = 12°
Θ_t = 135°

Fig. 1–10

QUESTIONS

1) Why is using a well name to identify a well a bad idea for a larger oil company?

2) Why is it easier to calculate the MLT than the TST?

3) In flat-lying stratigraphy, what would happen to the different thicknesses (TST, TVT, MLT, and TVDT) when a deviated well becomes horizontal?

4) What is the TVT of a unit that has a MLT of 40 m (131 ft), borehole dip of 36°, borehole direction of 89°, apparent horizon dip of 32°, and horizon dip direction of 78°?

5) What would the apparent vertical thickness of an interval (TST of 45 m [148 ft]) be if it were tilted 52°?)

2

Basic Well Logs and Log Signatures

BASIC WELL LOG TYPES

There are many types of geophysical well logs used today. The information derived from a geophysical log can be exact (e.g., using particular values to calculate *water saturations*) or interpreted (e.g., correlating stratigraphy using basic pattern recognition or identifying a section repeated by faulting). There are three main categories of logs: electric, radioactive, and structural (fig. 2–1). Electric and structural logs are typically run in *uncased* holes because the sensors need to be in contact with the borehole; whereas, radioactive logs can be run in either *cased* or uncased holes. With sub-meter drilling and sub-centimeter logging accuracies, identifying and drilling potential targets is becoming easier. The increased drilling accuracies have increased the demand for proper subsurface correlations by well log analysts and seismic interpreters. A fundamental understanding of each log is vitally important in understanding the subsurface geology. This chapter will cover basic well logs, what they measure, and common responses.

LOGS Electric Radioactive	Lithology	Porosity	Pressure	Hydrocarbons	Structure	Correlates	Mainly used to identify	Conditions for best use	Common units
Resistivity	X			X	X	Conductivity	Resistive vs nonresistive beds	Uncased hole	Ohm * m
SP	X					Permeablity	Permeable vs inpermeable beds	Uncased hole	mV
Gamma Ray	X					Radioactivity	Shaliness and organic content	Cased or uncased hole	API
Photoelectric	X					Mineralogy	Lithology	Cased or uncased hole	barns/e⁻
Neutron		X		X		Hydrogen content	Gas, porosity	Cased or uncased hole	% porosity
Density	X	X		X		Density	Gas, porosity	Uncased hole	g/cm³
Sonic		X	X	X		Velocity	Porosity, gas	Cased or uncased, gas-free	µs/ft
Dipmeter					X	Bed attitudes	Faults, unconformities, bedding	Uncased hole	Strike and dip
Caliper	X				X	Borehole size	Fractures	Uncased hole	in
Temperature				X	X	Well temperature	Faults, fractures, gas	Uncased hole	°F

Fig. 2–1. Chart of common logs and uses. SP = spontaneous potential.

Gamma Ray (GR)

A gamma ray (GR) log records the natural radioactivity of a formation in API (American Petroleum Institute) units. These units correlate with the gamma ray intensities measured by a *scintillation counter*. The radioactivity is the result of radioactive decay of mainly potassium (K), thorium (Th), and uranium (U).

Shale units will thus have high API values, while sandstone and limestone will have low API values. This is the most common log used for stratigraphic correlations (fig. 2–2); because it is widely run, it has good vertical resolution and is easy to interpret. Occasionally, high values correlate with borehole collapse features rather than stratigraphy, and other logs should be used to help identify potential *wash out* zones. The following are some typical values:

Coal: 20 API

Sandstone: 20 API

Shale: 75–200 API

Limestone: 20 API

Dolomite: 20 API

Salt: 0 API

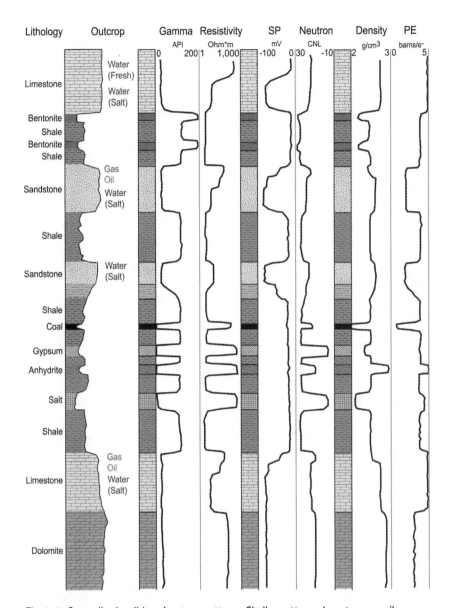

Fig. 2–2. Generalized well log signature patterns. Similar patterns do not necessarily indicate common depositional environments because different log types represent different borehole properties (i.e., lithology, density, and resistivity), and patterns are scale-dependent.

Resistivity

A resistivity log records the resistivity, or resistance, to the flow of electricity through a formation in Ohm meters (Ωm). Resistivity is the reciprocal of *conductivity* and is related to the porosity and the amount and kind of fluid present in the rock and borehole. The most important use of resistivity logs is in distinguishing hydrocarbons from water. A high-porosity or hydrocarbon-bearing formation has high resistivity; a low-porosity or saltwater formation has low resistivity. Numerous types of resistivity logs correspond mainly to the depth of the measurement (e.g., shallow, medium, and deep penetration). Shallow readings record the interface of the borehole and the drilling fluid, while deep readings correspond to *true* or *uninvaded* formation resistivity. Hydrocarbons may be present in a formation where deep resistivity is greater than shallow resistivity. Shales typically have low resistivity, and sandstones and carbonates have high resistivity.

Spontaneous Potential (SP)

A spontaneous potential (SP) log records the electrical current (in millivolts, mV) that arises due to salinity differences between a saltwater-based *drilling mud* and the fluid in a formation. This log is a good indicator of formation permeability and can distinguish shale from carbonates and sandstones. Another use of an SP log is the detection of hydrocarbons via a muted or diminished SP response. It is difficult to interpret SP logs because they are not good indicators of lithologic boundaries or general lithologies. In thin-bedded strata, the poor vertical resolution of the instrument will record a *bulk lithology* rather than separating the units. Shale typically has a value of 0 to -20 mV, and sandstones and carbonates typically have values between -20 and -80 mV.

Photoelectric (PE)

A photoelectric (PE) log records gamma radiation transmitted from a formation after being bombarded by *photons*. The PE factor is measured in barns per electron (barns/e-). The amount of photoelectric absorption within the formation depends on the mineralogy of the formation and, therefore, is an excellent indicator of lithology. This is one of the few geophysical logs that can readily distinguish limestone from dolomite because of the almost 2:1 PE response. Following are some typical values (adapted from Serra 1984):

Coal: 0.17 (barns/e-)

Sandstone: 1.81 (barns/e-)

Dolomite: 3.14 (barns/e-)

Shale: 4.0 (barns/e-)

Limestone: 5.08 (barns/e-)

Salt: 4.7 (barns/e-)

Neutron

A neutron log measures a formation's porosity based on the quantity of hydrogen present in the formation. Fluid within a formation as opposed to gas can be identified because water and oil have approximately the same hydrogen content, whereas gas has less hydrogen per equal volume. Hence, if gas is present in a formation, the *compensated neutron log* (CNL) will underestimate porosity. Salt units have characteristically low neutron porosity and bulk density readings.

Density

A density log measures the porosity of a formation based on the assumed density of the formation and drill fluid in grams per cubic centimeter (g/cm^3). The standard porosity calculation (grain density − measured bulk density)/ (grain density − drilling fluid density) will overestimate the porosity of a gas-filled formation because the measured bulk density will be lower. Hence, when overestimated porosity values (from a density log) are cross-plotted with the underestimated porosity values (from a neutron log), the crossover is an indication of gas in a formation called the *gas effect*. Shale, coal, and bentonite beds commonly have low densities and sandstones, and carbonates generally have higher densities. Some typical values are the following (from Serra 1984):

Grain density of shale: 2.4–2.6 g/cm^3

Grain density of sandstone: 2.65 g/cm^3

Grain density of limestone: 2.71 g/cm^3

Grain density of dolomite: 2.87 g/cm^3

Grain density of salt: 2.03 g/cm^3

Density of drilling mud: 1–1.1 g/cm^3

Density of water: 1.0 g/cm^3

Density of crude oil: 0.8–1.0 g/cm^3

Density of natural gas: 0.7 g/cm^3

Dipmeter

A dipmeter log measures the strike and dip of the strata encountered in the wellbore in degrees. A dipmeter log is usually represented by a *tadpole plot*. The plot consists of a series of circles (tadpole heads) placed at the depth where the measurement was taken. Dip is illustrated by the location of the circle on the horizontal axis (0° to 90°) and the strike is illustrated by the azimuthal direction of a line coming off the circle (the tadpole tail). These logs can easily confirm structural and sedimentary features (i.e., faults and cross-bedding), but are not commonly run.

Caliper

A caliper log records the diameter of the borehole in inches (in.). Caliper logs can be used as a crude lithologic indicator. Shale, coal, and bentonites tend to cave into the borehole and produce an increased well diameter. Well-*indurated* sandstones and carbonates tend not to cave, and subsequently not produce a deviation from the well diameter. A decreased well diameter frequently indicates drilling mud has become caked to the borehole or the presence of *montmorillonite*.

Sonic (Acoustic)

A sonic (acoustic) log records the speed of sound transmitted through a formation in microseconds per foot (μs/ft). The speed at which the rock transmits sound energy is related to the formation's porosity. The lithology of a formation must be known to accurately calculate porosity. A sonic log is a good indicator of density and presence of gas. Measurements will be lower and have locally irregular signatures where gas is present because the gas has a slower transmit time. The following are some typical compression velocities for consolidated rocks (from Hearst et al. 2000; Asquith and Krygowski 2004):

Shale: 62–167 μs/ft

Sandstone: 55.5 μs/ft

Limestone: 47.5 μs/ft

Dolomite: 43.5 μs/ft

Temperature

A temperature log records the temperature present in the wellbore in degrees Fahrenheit (°F). Temperature typically increases with depth at a constant rate (the *geothermal gradient*). Geothermal gradients vary from place to place depending on numerous tectonic, stratigraphic, and geochemical factors. The overall gradient is not as important in identifying structures (i.e., faults and fractures) as the temperature deviations, or kicks. A localized temperature decrease is indicative of fluid or gas emanating from a fault or localized fracture system, and a broad temperature decrease is indicative of fluid or gas emanating from a pervasive fracture system. The decrease in temperature is typically caused by expansion and subsequent cooling of gas invading the wellbore.

Log Signature Patterns

Pattern recognition plays a crucial role in correlating subsurface intervals. Lithologies and *facies* can change vertically and laterally, but because the shape of the signature or pattern is usually similar, it is possible to correlate units across large areas. There are two categories of patterns (symmetric and asymmetric) and six types of patterns (irregular, flat, funnel, bell, cylindrical, and bow). Symmetric patterns are trends that have a horizontal line of symmetry (flat, *cylindrical*, and *bow*), whereas asymmetric patterns (irregular, *funnel*, and *bell*) do not have horizontal symmetry. Irregular patterns (fig. 2–3) (i.e., trends that have sharp boundaries with similar magnitudes and have a saw blade appearance) indicate an *aggradational* or fluctuating depositional environment. Flat patterns (i.e., trends that do not have sharp boundaries and have a consistent magnitude) indicate steady-state depositional environments. Funnel patterns (i.e., signatures that gradually decrease) are indicative of coarsening or fining upward sections (depending on the log type). Bell patterns (i.e., signatures that gradually increase) are also indicative of coarsening or fining upward sections (depending on the log type). Cylindrical patterns are flat trends bounded by sharp edges. This pattern indicates a steady-state depositional environment. Bow patterns that have a gradual decrease, then increase in the signature, are indicative of coarsening or fining upward sections.

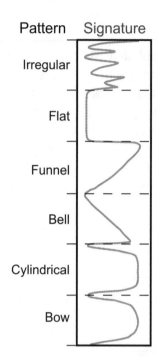

Fig. 2–3. Generalized well log signature patterns. Similar patterns do not necessarily indicate common depositional environments because different log types represent different borehole properties (i.e., lithology, density, and resistivity) and patterns are scale-dependent.

EXERCISE

2–1) Identify the lithologies in the following well logs (figs. 2–4 through 2–9).

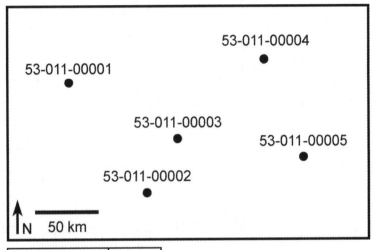

API #	Elv. (m)
53-011-00001	432
53-011-00002	430
53-011-00003	422
53-011-00004	429
53-011-00005	436

Fig. 2–4

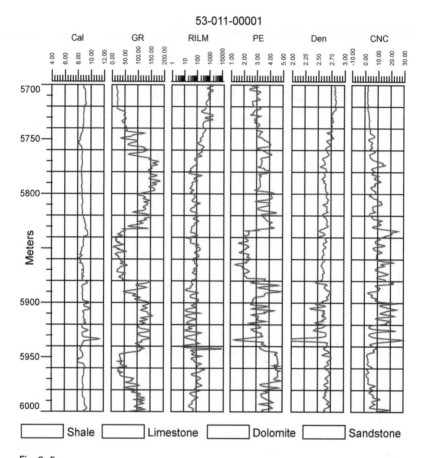

Fig. 2–5

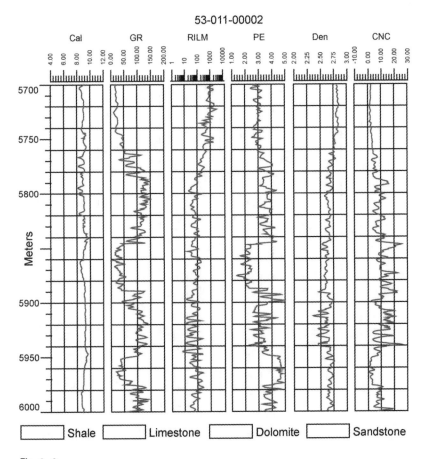

Fig. 2–6

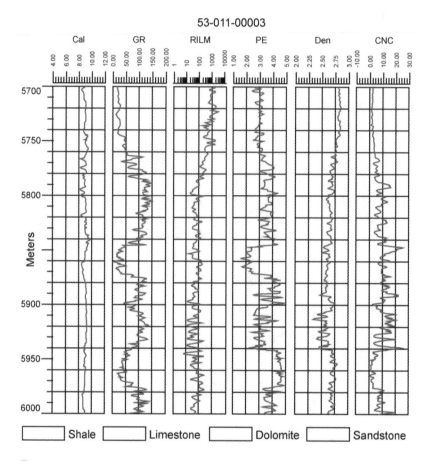

Fig. 2–7

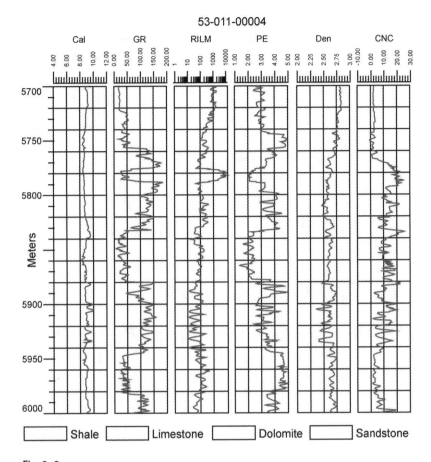

Fig. 2-8

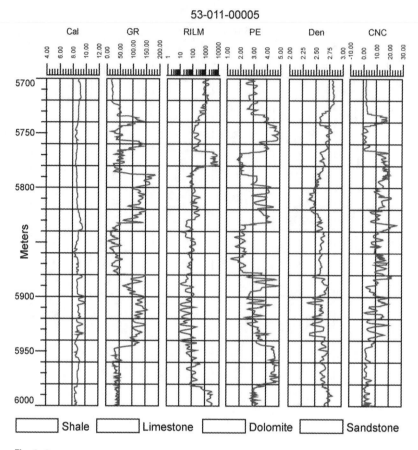

Fig. 2–9

QUESTIONS

1) What is the gas effect?

2) Why do electric logs have to be run in an uncased hole?

3) What are two types of asymmetric patterns?

4) What lithology is probably present if the gamma ray, neutron porosity, and density values are low?

5) What common sedimentary rock type has the fastest compressional velocity? (Be careful to look at the units.)

3

INTRODUCTION TO SUBSURFACE MAPS AND CONTOURING

In order to interpret subsurface geology, it is necessary to become familiar with different types of subsurface stratigraphic maps. The main types are structure contour, *isopach*, *trend surface*, and *trend surface residual anomaly maps*. These maps can be constructed from almost any type of XYZ dataset (well, *potential field*, or *seismic reflection* data), but this book will only cover well data. A structure contour map is a map of a particular horizon commonly associated with a prominent *marker bed*, horizon, or interval; it is used to locate *structural highs* and structural trends. An isopach map is a map of the true stratigraphic thickness between two horizons. These maps are useful to differentiate between structural and stratigraphic traps, identify depositional changes, and determine possible locations to drill for hydrocarbons and oil. A trend surface map is a best-fit surface that estimates the overall shape of a surface. Higher-order trend surfaces better represent the data but will be less effective in estimating trends because they more closely resemble a structure contour map. Trend surface maps can also be used to calculate residual anomalies and identify regional depositional changes. In chapter 8, basic 3-point problems will be utilized to quickly approximate a trend surface.

CONTOURING

Contouring is an integral step in representing data on a map. Basic contouring laws must be followed, but there is stylistic liberty where there is limited information or if regional geology indicates probable patterns. A contour map has data on it separated by *isolines*, which are lines of equal value. The name of an isoline is dependent on the data being contoured (e.g., temperature = *isotherm*; time = *isochron*; stratigraphic thickness = isopach; true vertical thickness = *isochore*; elevation = *contour line*). Contouring laws are straightforward: 1) a contour line cannot merge with other contour lines; 2) repeated contours indicate a slope reversal; 3) hachured contour lines indicate depressions; 4) contour lines cannot cross each other (except for overhanging surfaces); 5) the contour interval must remain constant; 6) every fifth contour is labeled; 7) all values between two contour lines must be within the contour range; and 8) each contour is a continuous line that must be closed within the map, project outside the map, or intersect a *fault break*. A common weather map (fig. 3–1) is a good example of a basic contour map. Since this type of map may have differing projections, a latitude-longitude grid typically will replace a scale bar. Fault breaks indicate a thrust or normal fault, and data are contoured independently on either side of the break (fig. 3–2).

To construct a contour map, first accurately plot the desired data on a base map, and add a scale bar and north arrow. Then determine the maximum and minimum values and choose a *contour interval* that would allow a desirable number of contours to be drawn on the map. Too many contour lines will clutter the map and too few contour lines will not adequately portray data patterns. Next, draw contour lines through and around the data until all the data falls properly between two contour lines. Make sure contour lines pass closer to data with similar values and farther away from data with considerably different values. Sometimes it is necessary to draw a circular contour line around a data point, but this may indicate an invalid data point.

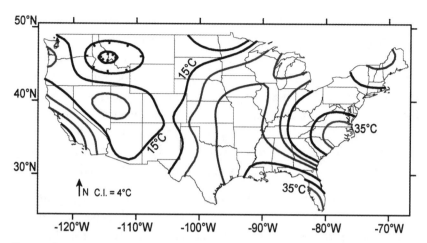

Fig. 3–1. A temperature map for the continental U.S. (C.I. = contour interval).

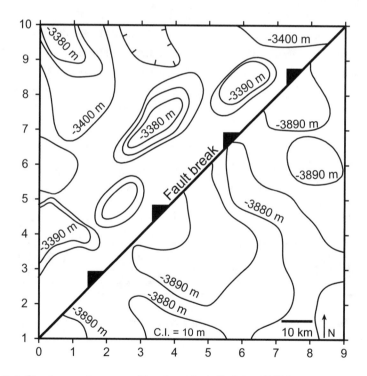

Fig. 3–2. Structure contour map with a thrust fault (fault break) (C.I. = contour interval).

Finish drawing all the contour lines on the map and note the contour interval on the map. It should become evident that closer contour lines (fig. 3–3) represent greater gradients and widely spaced contours represent less variance in the data. A contour map is valid as long as all the data have been honored and no contouring rules have been violated. Even though a map honors all the data, it does not mean the map is accurate. Various interpretation and contouring patterns are possible for any given dataset. Less data plotted on a map means there will be a greater number of possible representations. More data or background information may indicate that the map should be altered or redrawn in a different style.

Using computers to contour data is extremely common because they are very efficient and accurate. It is also important to remember that computers are not biased. They operate as they are programmed, regardless whether the operation is appropriate or the resulting map is valid. A user must be able to properly choose what contouring algorithm is most appropriate to properly display the data and whether any contouring breaks or bounding requirements are needed. Therefore, a *black box* approach to the graphical display of data should be avoided. Contouring data by hand or using a computer should be an iterative process, in that different maps should be constructed and the best or most useful map should be chosen. Having a solid background in the regional geology will help in selecting the most suitable map.

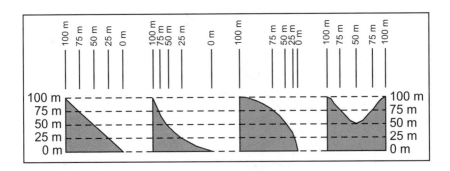

Fig. 3–3. Two-dimensional views of differently spaced contours. Closer contours represent larger gradients.

EXERCISES

3–1) Contour the following data and complete the two profiles (fig. 3–4).

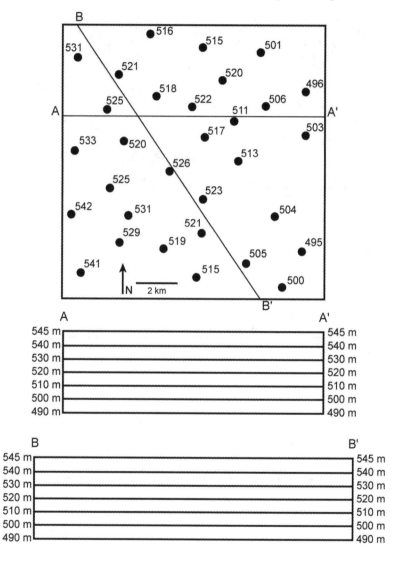

Fig. 3–4

3–2) Draw three structure contour maps to illustrate what the *décollement-related* anticlines in the cross section may look like in map view (fig. 3–5). Assume the thrust faults and associated anticlines are striking perpendicular to the page. The contour values are not needed.

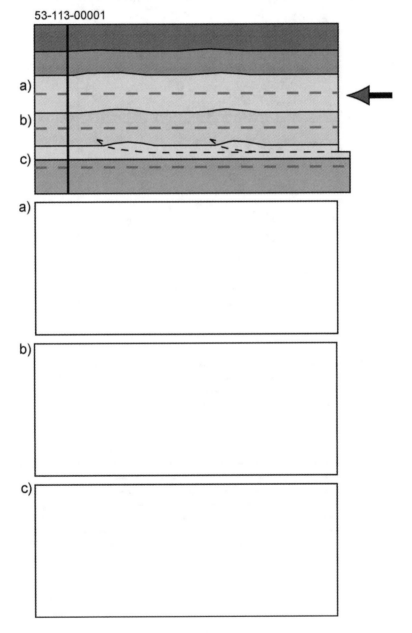

Fig. 3–5

3–3) Identify the lithologies in the following well log (fig. 3–6).

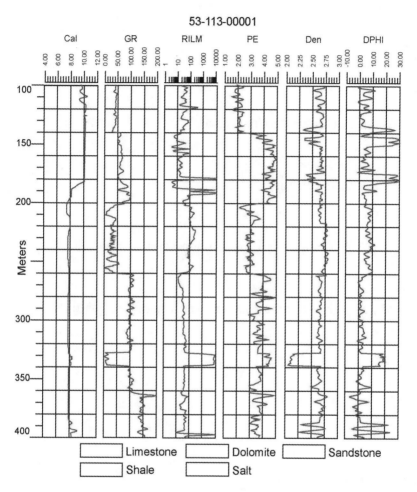

Fig. 3–6

QUESTIONS

1) List three types of subsurface maps.

2) What does a hachured isoline indicate?

3) What is an isopach map?

4) In fig. 3–1, what is the lowest isotherm shown on the map?

5) Why are higher-order trend surface maps commonly of limited value?

STRUCTURAL AND STRATIGRAPHIC INTERPRETATIONS

PICKING METHODOLOGY

In order to map the subsurface, it is necessary to learn how to effectively pick key or mappable horizons, contour the data, and visualize the data in three dimensions (fig. 4–1). Picking mappable horizons can be difficult because of stratigraphic changes (e.g., facies changes), structural features (e.g., faults), or vertical and lateral discontinuities (e.g., unconformities). Ideal marker beds are units or lithofacies that are widespread, distinctive, and lithologically *homogeneous*. Stratigraphic contacts typically make good markers (fig. 4–2), although the contact of an unconformity with similar lithologies above and below the contact may be difficult to identify. A sandstone channel may be important for hydrocarbon exploration, but overall, is not a very good marker bed. A *bentonite* or ash bed is an excellent marker, but it may not be very important in hydrocarbon exploration. In order to start searching for a desired target (i.e., oil-filled sandstone, a high-grade ore body, an *aquifer*, or a hydrocarbon *trap*), it is necessary to understand the regional subsurface framework.

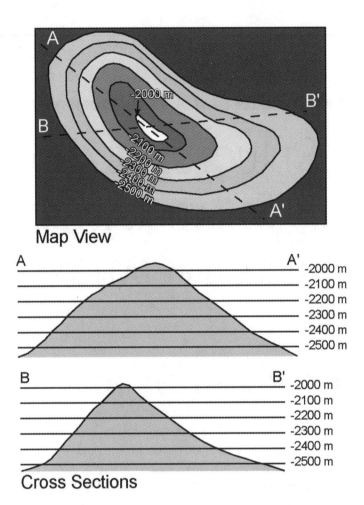

Map View

Cross Sections

Fig. 4–1. Cross sections through a structural high. Note the asymmetry of the feature.

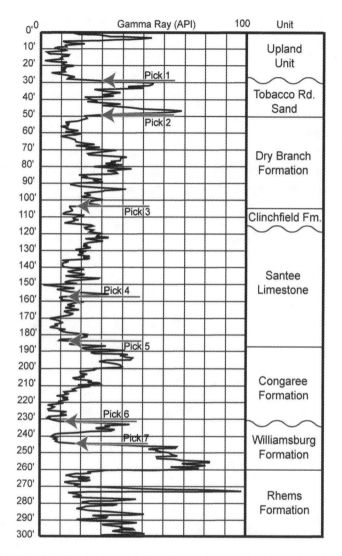

Fig. 4–2. Type log with multiple picks from Savannah, Georgia. Certain picks (i.e., picks 4 and 5) do not correspond to stratigraphic tops and cannot be regionally correlated.

STRUCTURAL INTERPRETATIONS

Common structural features that can be identified in subsurface maps and geophysical logs are folds (fig. 4–3) and faults (figs. 3–2 and 4–4). Faults may be present when there are repeated sections on a log (thrust faults), a known section is missing (normal faults), closely spaced contours are present in a structure contour map, or when there is a pronounced thickening of an interval in an isopach map. Folds may be present when (1) inverted sections are present on a log (*recumbent* or overturned folds), (2) a pronounced thickening is juxtaposed between two thinned regions not associated with a fault in an isopach map (due to material flow into the *hinge* of a fold), or (3) a high is juxtaposed between two *structural lows* in a structure contour map not associated with thickening. If seismic data were available, it would be easy to identify such structures as long as (1) the structures were above seismic resolution, (2) bedding was not significantly inclined, and (3) the fault was not parallel to bedding.

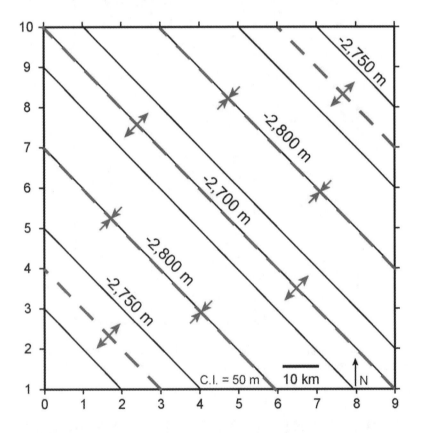

Fig. 4–3. Structure contour map showing multiple folds. Assuming there is no stratigraphic or structural thickening, the corresponding isopach map would not have any linear pattern.

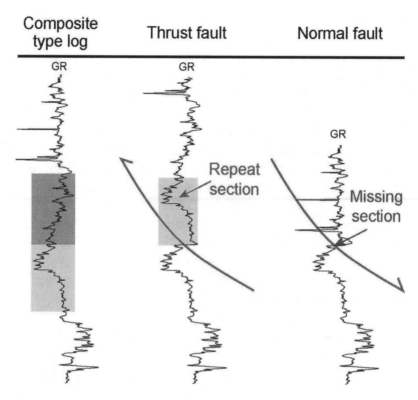

Fig. 4–4. A composite type well log derived from two nearby wells that are faulted. The light gray section in the composite log indicates the interval repeated in the thrust fault log and the dark gray section indicates the interval missing in the normal fault log. This is a composite type log because it was generated by adding the two wells. A type log must be from a well that has a complete and unfaulted section.

STRATIGRAPHIC INTERPRETATIONS

Stratigraphic features are often more difficult to delineate and identify in the subsurface. Commonly, these features need to be interpreted using both structure contour maps and isopach maps. To help recognize bedding attitudes, dipmeter data or *formation micro-scanner* (FMS) logs are required. A FMS log creates a 360° synthetic representation of the borehole wall using small resistivity variations. This log can aid in identifying bedding planes, fractures, faults, *foliations*, and possibly lithologies. Other types of maps, such as trend surface maps, can identify regional dip of horizon, which may indicate a depositional *source* or vertical movement related to tectonism, but an isopach map is the primary map used to make stratigraphic interpretations. The map is the representation of thickness between two picks. Isopach map patterns, like any map patterns, are non-unique. These maps, however, can rule out structural features and quickly identify thickening trends related to deposition. Linear trends are probably related to large-scale depositional patterns, and irregular (fig. 4–5) or *bull's-eye patterns* could indicate reefs or dissolution features. A cross section through the target interval and lithologic correlations will help constrain an inferred geologic model. Lateral facies changes indicate large-scale depositional patterns, whereas localized thickened or facies changes associated with less well-defined trends, point to possible reefs, dissolution, or *dewatering* features. Sometimes the lithologies present in the wells will rule out these interpretations and force the interpreter to alter the existing model. For example, an interval that is dominantly sandstone is not likely to have major reef trends within it, and an interval composed mostly of *turbiditic* shales will not have large-scale dissolution features.

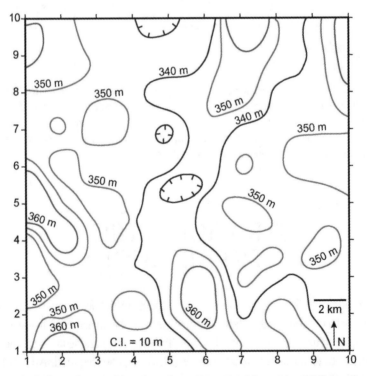

Fig. 4–5. An isopach map with an irregular pattern except for a minor NNE trending feature near the middle of the map that could be related to a structural feature.

EXERCISES

4–1) Identify the lithologies in the following wells and correlate the wells on the provided cross section (figs. 4–6 through 4–13).

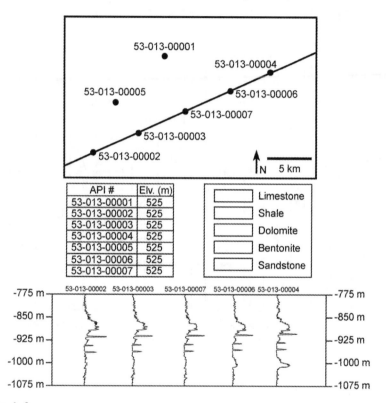

Fig. 4–6

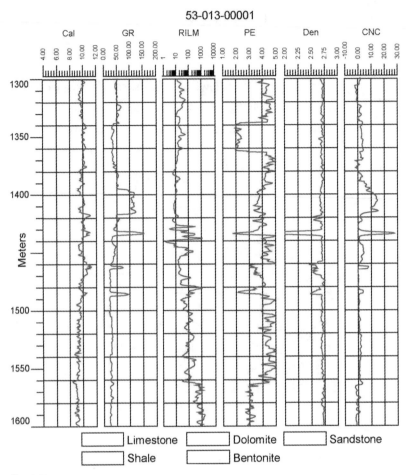

Fig. 4–7

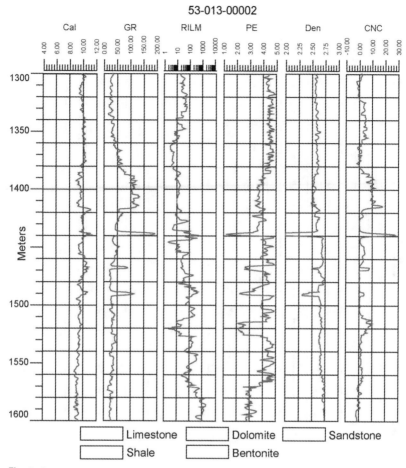

Fig. 4–8

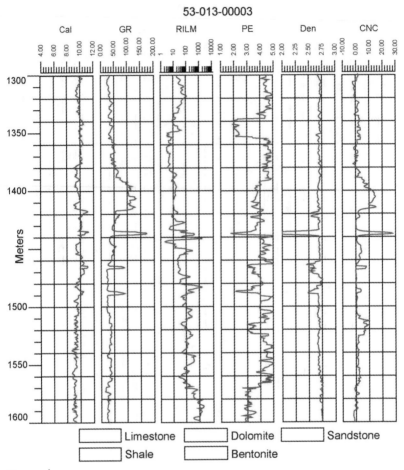

Fig. 4–9

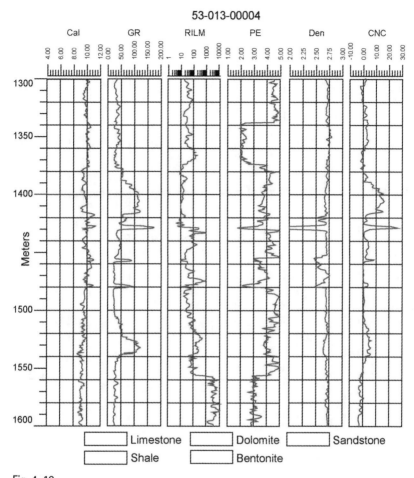

Fig. 4–10

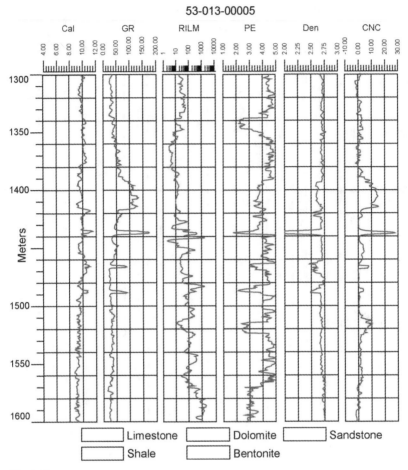

Fig. 4–11

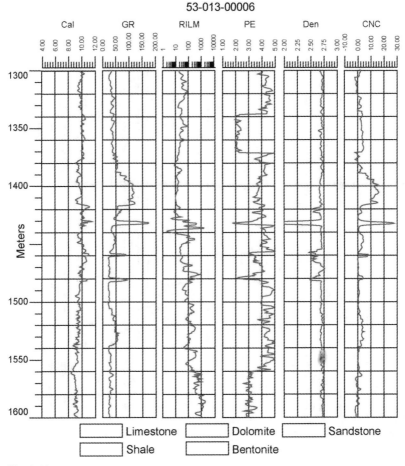

Fig. 4-12

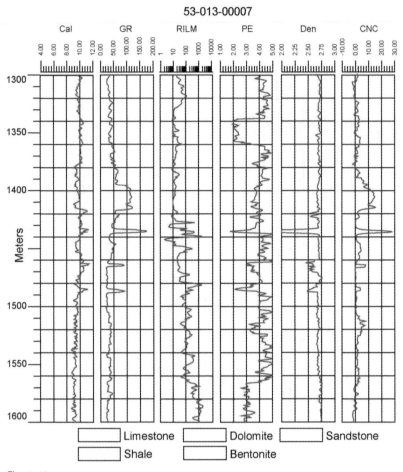

Fig. 4–13

QUESTIONS

1) Why must multiple types of subsurface maps be used to identify a trend or target?

2) What makes a good marker bed?

3) How is a composite type log different from a type log?

4) Why are normal faults more difficult to identify than thrust faults in a well log?

5) Looking at your finished cross section in Exercise 4–1, explain why gamma ray correlations are not always reliable.

5

STRUCTURE
CONTOUR MAPS

A structure contour map is the most common map type used to represent and investigate the subsurface. Structure contour maps are made to observe trends (fig. 5–1) and are constructed from pick data relating to the top of a marker bed, key interval, fluid contact, fault surface, seismic reflection surface, or ore horizon. The data represented in the maps are basic location (X/Y) and depth data referenced to a datum. Referencing a data point or subsurface pick is achieved by subtracting the log datum elevation from the interpreted pick elevation. The log datum is commonly the K.B. or G.L.; this information can be found in the well header or ticket. The elevations are standardized to sea level and the data should be properly labeled, such as meters below sea level (MBSL) or feet below sea level (FBSL).

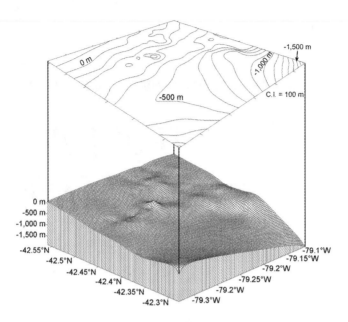

Fig. 5–1. Three–dimensional, wireline mesh representation of a structure contour map having an east–west trend in the northern part of the map and a northeastern trend in the southern portion of the map.

A structure contour map can be thought of as a subsurface elevation map. If all the overburden were stripped away, the resulting surface would be a topographic map. A structure contour map is therefore particularly useful in identifying and delineating structural trends or *topographic trends* probably related to structural or stratigraphic features. In order to properly delineate and identify structures, it is necessary to use multiple structure contour maps (fig. 5–2) and other map types because, like geophysical data, maps represent *non-unique* data. In fig. 5–2, it is evident that the three maps (stacked or spatially referenced) represent an angular unconformity. Without the use of multiple maps however, it would be impossible to differentiate the stratigraphic relationships between the horizons. In areas with large-scale structures or stratigraphic features, it is very easy to identify potential economic targets, but it is very difficult to identify these targets on smaller-scale maps or in regions with reactivated or *overprinted* structures.

This introduces the concept of scale-dependent maps. The choice of map scale and contour interval are very important. A small-scale fault (<5 m [<16.4 ft] displacement and a fault trace or length of 1,000 m [3,281 ft]) will not be evident in a map if the contour interval is 100 m (328 ft) or if the map scale is 1:1,000,000. The opposite scenario is also a problem, where the target comprises the entire map and the contour interval is too small to fully comprehend the target size. This represents the common problem of *not seeing the forest for the trees*. To address this problem, the standard practice is to start investigating larger-scale or regional problems and then focus or zoom-in on individual areas that are regions of interest. Again, it is critical to use multiple types of maps and data whenever possible. This will help constrain the number of possible solutions and limit the number of plausible geological models.

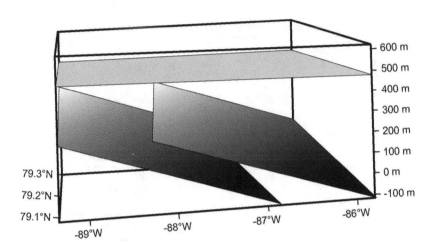

Fig. 5–2. Three-stacked structure contour maps illustrating an angular unconformity.

EXERCISES

5–1) Identify the lithologies in the following wells and calculate the depth to
the top of the salt and bentonite units (figs. 5–3 through 5–15). Create
structure contour maps for the tops of bentonite and salt; remember to
indicate the contour interval and complete the cross section A-A in the
space provided. Use the outcrops as contouring breaks.

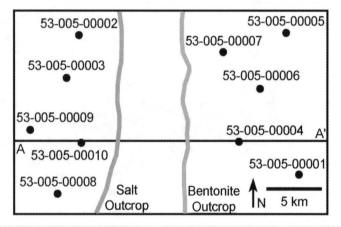

API #	Elv. (m)	Salt (m)	MBSL (m)	Bentonite (m)	MBSL (m)
53-005-00001	926				
53-005-00002	1028				
53-005-00003	1027				
53-005-00004	954				
53-005-00005	903				
53-005-00006	930				
53-005-00007	945				
53-005-00008	1009				
53-005-00009	1025				
53-005-00010	1011				

Fig. 5–3

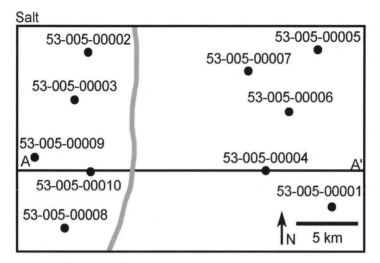

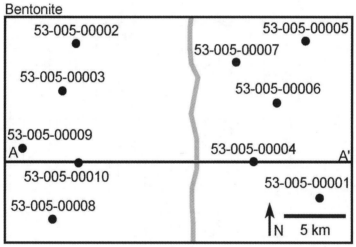

Fig. 5–4

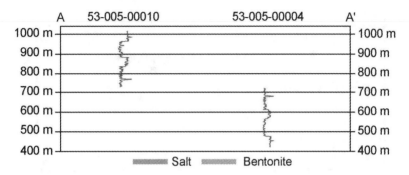

Fig. 5-5

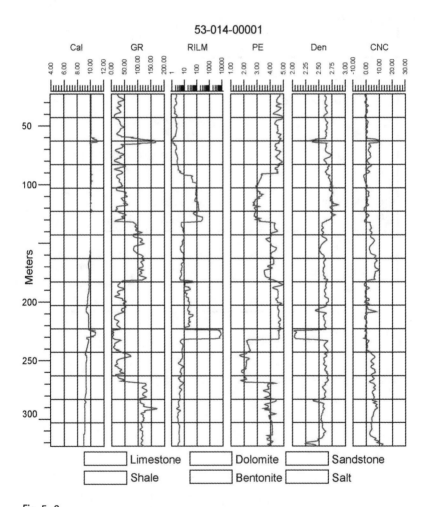

Fig. 5-6

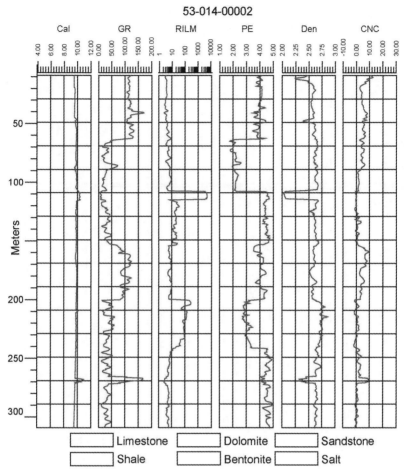

Fig. 5–7

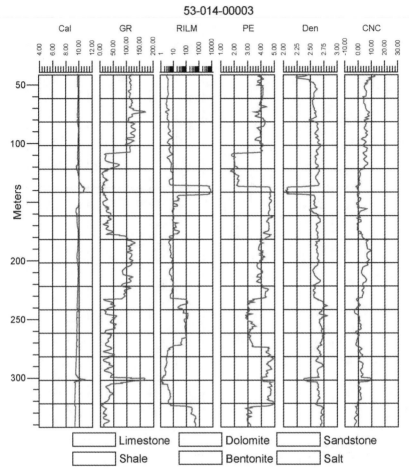

Fig. 5-8

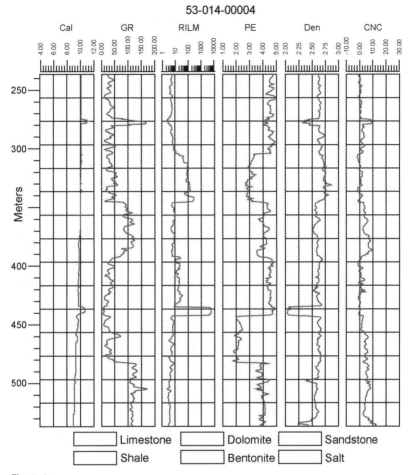

Fig. 5–9

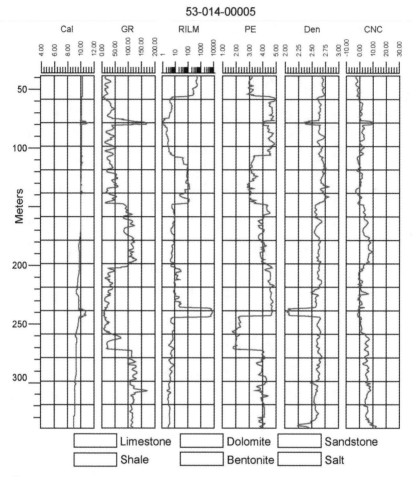

Fig. 5–10

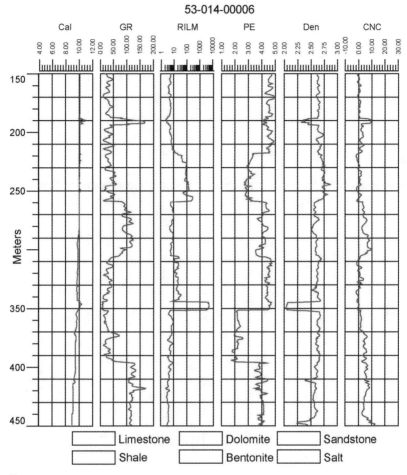

Fig. 5–11

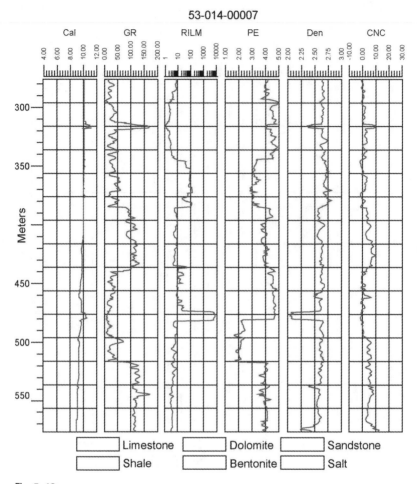

Fig. 5–12

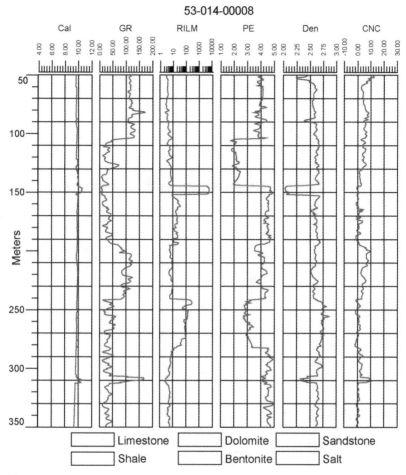

Fig. 5-13

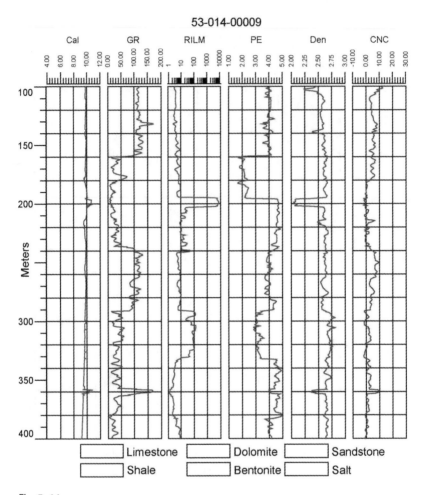

Fig. 5–14

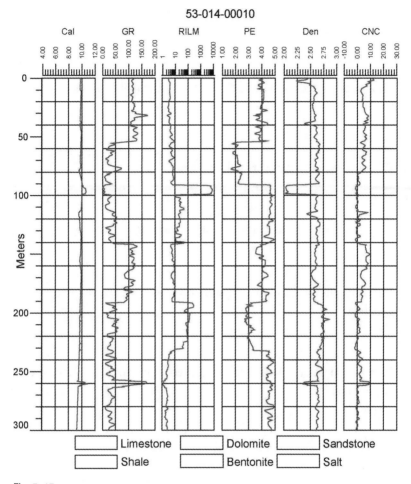

Fig. 5-15

QUESTIONS

1) What types of geological features do structure contour maps usually represent?

2) Where is the well log elevation datum found?

3) Why is choosing the proper map scale important?

4) Why are multiple structure contour maps more useful when used together?

5) How do you find the XYZ information needed to construct a structure contour map?

6

THICKNESS MAPS

Thickness maps are another type of standard map used to explore for natural resources and to interpret stratigraphic relationships. Thickness maps are a graphical representation of variations in stratigraphic thickness. There are many different types of thickness (TST, TVT, MLT, TVDT, and net pay), and when each type of thickness is mapped, the maps have different names. A map of TST is called an isopach map, a TVT map is referred to as an isochore map, and a *net pay* map is conveniently called a net pay map. Net pay is the thickness of a potential *reservoir* capable of producing water or hydrocarbons.

A prevalent error is to refer to all thickness maps as isopach maps. This can lead to considerable misinterpretations (fig. 6–1). For example, a simple fold has a constant TST, but the TVT will thicken in the *limbs* of the fold. If an isochore map were constructed for one of the folded intervals and labeled an isopach map, the structure might be interpreted as two faults rather than a simple fold. MLT and TVDT data are commonly not contoured because they are not very useful when interpreting structural and stratigraphic features. The thicknesses are important however in reservoir calculations and production estimates.

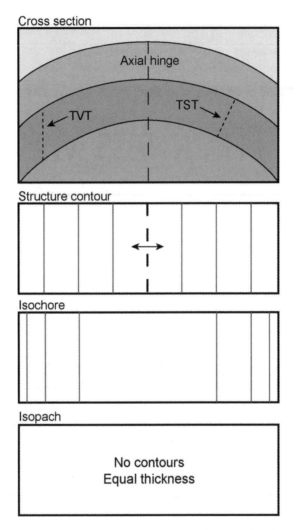

Fig. 6–1. Structure contour, isochore, and isopach maps of a simple fold. Note that confusing an isochore and isopach map may lead to a misinterpretation of the structure.

To make an isopach or isochore map, it is necessary to measure the TVT or TST between two marker units or beds. The resulting thicknesses are plotted on a well location map and the data are contoured normally. Linear thickening trends are commonly the result of thrust faulting or depositional features, such as a marine channel or bar sand. An isolated thickening trend may represent a *karst* feature, an isolated *pinnacle reef,* a reef complex, a *salt dome* (fig. 6–2), or an erroneous data point. A prominent, oriented thickening may indicate a sediment source direction; sediments should thicken toward the source region. Using other types of maps will help constrain these interpretations.

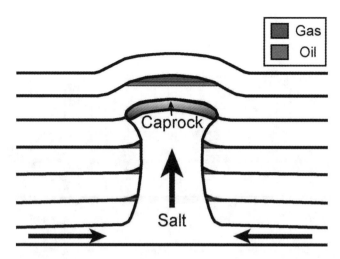

Fig. 6–2. Cross section of a salt dome trap

Net pay maps are an important type of thickness map in the exploration of hydrocarbons or water *reserves*. Reservoirs or aquifers are *heterogeneous*, and as a result, the entire reservoir interval is not generally productive. Porosity and permeability vary laterally and vertically throughout any reservoir. It is necessary to get an estimate of how much of a given interval or reservoir is potentially a *pay zone*. A minimum porosity is usually defined as a *cutoff*, and will depend on economics, the field, and the exploration company. Four percent is a common cutoff because less than 4% usually means the reservoir has limited permeability and is thus incapable of effectively draining the limited amount of hydrocarbons it may contain. Any section within a potential reservoir interval having porosity higher than the cutoff is considered a net pay (fig. 6–3). The amount of net pay in a given interval is then summed and contoured as any standard map.

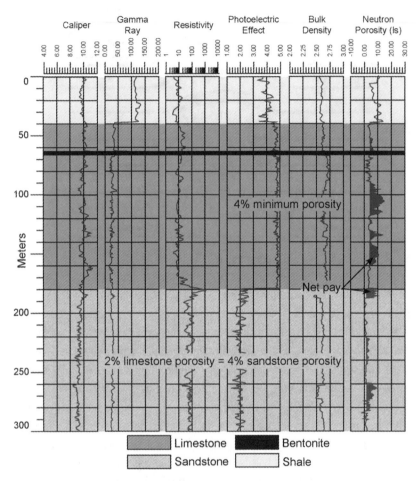

Fig. 6–3. A well log showing the net pay within a limestone and sandstone reservoir using a 4% cutoff neutron porosity. Notice the sandstone porosity had to be adjusted because the neutron log was run on a limestone matrix. Choosing an appropriate conversion chart will depend on which porosity logs are run, drilling fluid density, and an individual's preference.

Although computer programs can quickly calculate net pay values, they can be made manually relatively easily. To calculate a net pay value for a particular interval, the lithologies must be properly interpreted. Then the porosity values must be adjusted based on the log and the lithology (fig. 6–4). If the density porosity of a particular log has been run on a limestone matrix and the porosity of a sandstone section has been plotted as 15%, the porosity will have to be adjusted to 12% to compensate for the difference between the densities of sandstone and limestone. There are many types of porosity conversion charts because porosity can be derived from different logs (neutron, sonic, or density), and drilling fluids have different densities. Choosing an appropriate conversion chart will depend on which types of logs are run, drilling fluid density, and an individual's preference. After all the adjusted porosity data have been calculated, the values in an interval can be totaled and a *porosity-foot* and average porosity can be calculated. A porosity-foot is the total amount of void space in a given interval and the average porosity is the porosity-foot valued divided by the thickness of the interval.

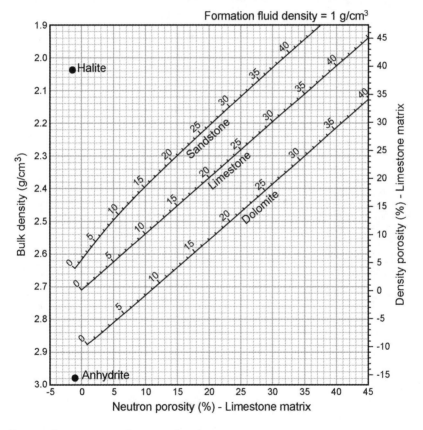

Fig. 6-4. A common porosity conversion chart

EXERCISES

6–1) Identify the lithologies in the following wells and calculate the depth to the top of the coal (pick 1) and limestone (pick 2) units (figs 6–5 through 6–17). Create structure contour maps for the coal unit and an isochore map for the interval between picks 1 and 2. Complete the cross section A-A in the space provided using structure contour data and lithologic correlations.

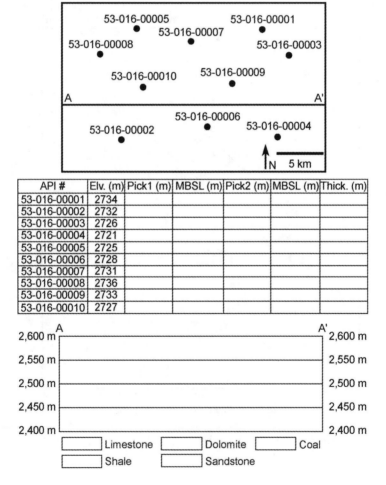

API #	Elv. (m)	Pick1 (m)	MBSL (m)	Pick2 (m)	MBSL (m)	Thick. (m)
53-016-00001	2734					
53-016-00002	2732					
53-016-00003	2726					
53-016-00004	2721					
53-016-00005	2725					
53-016-00006	2728					
53-016-00007	2731					
53-016-00008	2736					
53-016-00009	2733					
53-016-00010	2727					

Fig. 6–5

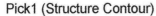

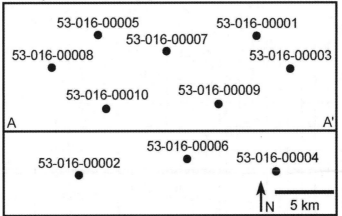

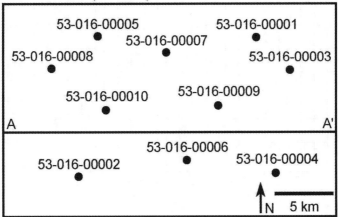

Fig. 6–6

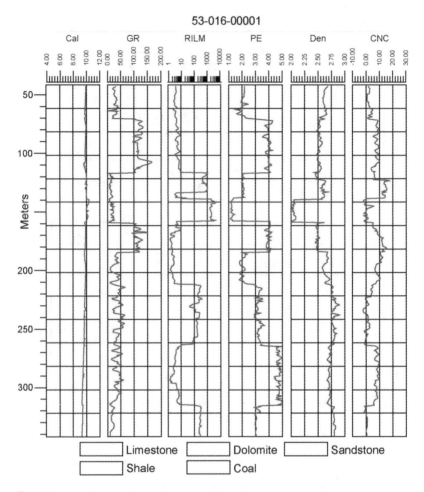

Fig. 6–7

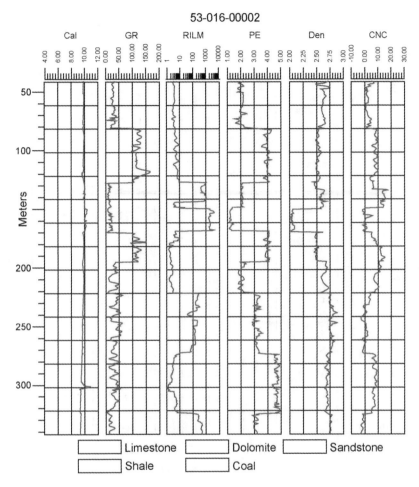

Fig. 6–8

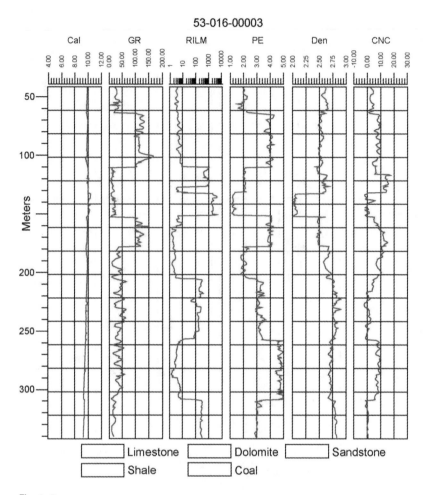

Fig. 6–9

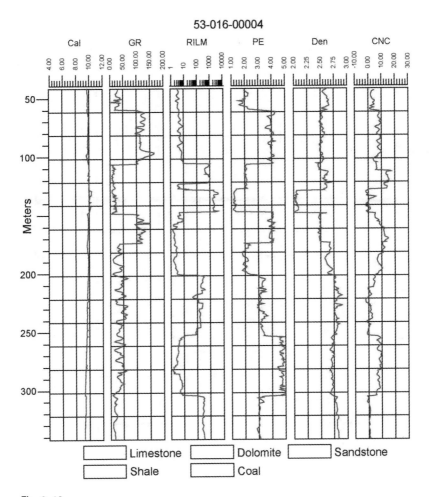

Fig. 6–10

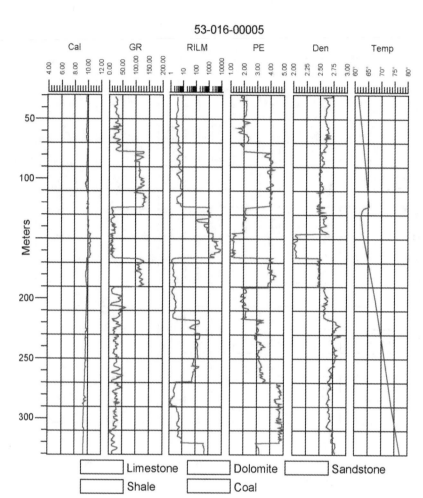

Fig. 6–11

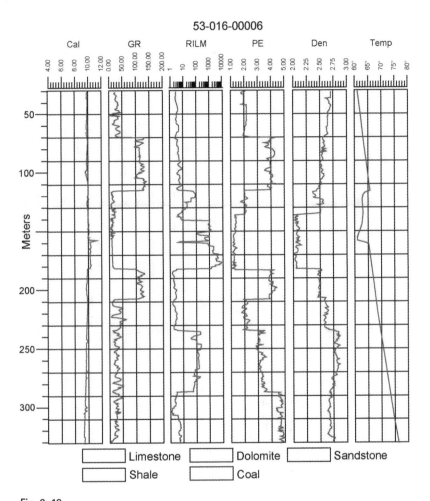

Fig. 6–12

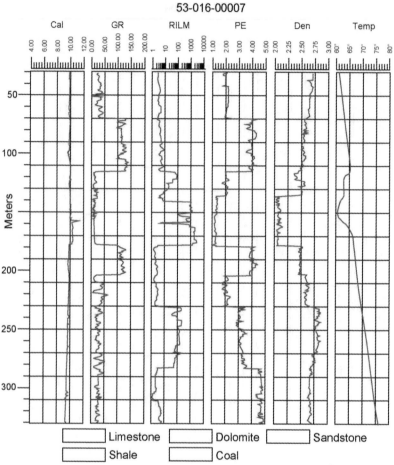

Fig. 6–13

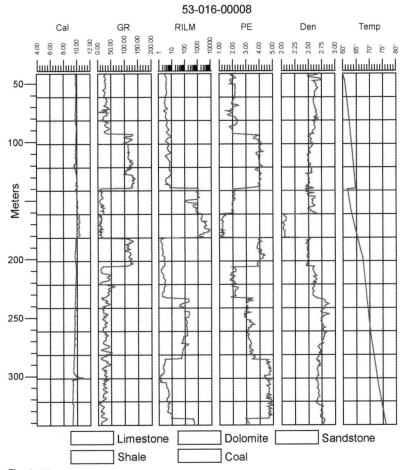

Fig. 6–14

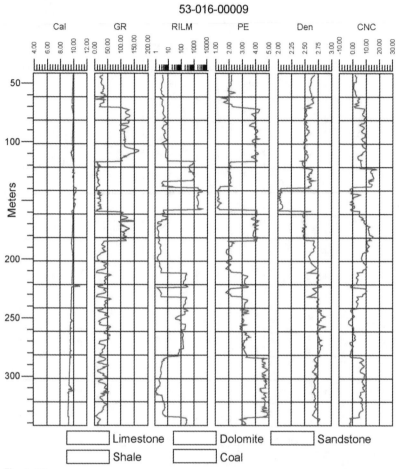

Fig. 6–15

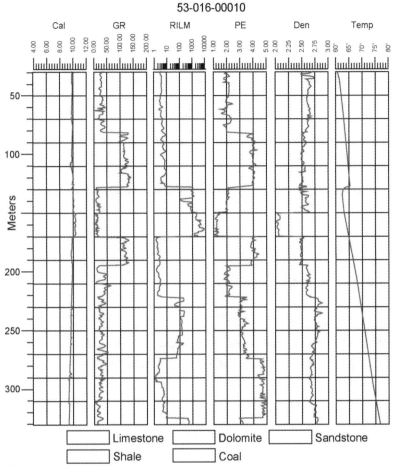

Fig. 6-16

QUESTIONS

1) What types of features are thickness maps able to identify or delineate?

2) What is the difference between an isopach and isochore map?

3) What is net pay?

4) What will cause neutron porosity to be erroneously low and what effect will this have on net pay estimates?

5) What is the net pay in the uppermost sandstone interval? Use the neutron density log (limestone matrix) and 5% porosity as the minimum cutoff for sandstone.

7

FACIES MAPS

A facies or *lithofacies* is a section of rock having similar characteristics or general appearance. A particular facies has an inferred depositional environment(s). Lithofacies are almost always *diachronous*, meaning correlatable facies vary in age and cross *biozones* and *time lines*. Facies maps, therefore, represent paleodepositional environments just as structure contour maps usually represent a lithofacies surface. A facies map can facilitate identifying source direction, changes in sea level, paleodepositional environments, possible permeability and porosity, and potential exploration trends.

A single borehole, if it represents an unfaulted and continuous section with no unconformities, can be used to interpret local or possibly regional changes in sea level, the amount of accommodation space, or sediment supply. *Walther's law* is a simple, yet elegant way to visualize *regressions* and *transgressions* probably caused by changes in sea level, subsidence or uplift rates, or rates of seafloor spreading. Walther's law states coeval facies are deposited laterally and facies stacking order reflects changes in depositional environments through time (fig. 7–1). This implies laterally adjacent facies will overlie the current facies if the depositional environment changes. For example, a stratigraphic section with coarse-grained sandstone at the base, that progressively grades upward into a siltstone and ultimately a shale unit at the top, reflects a transgression. If this section is turned sideways, it is possible to interpret the paleodepositional environments, facies source direction, and possible provenance.

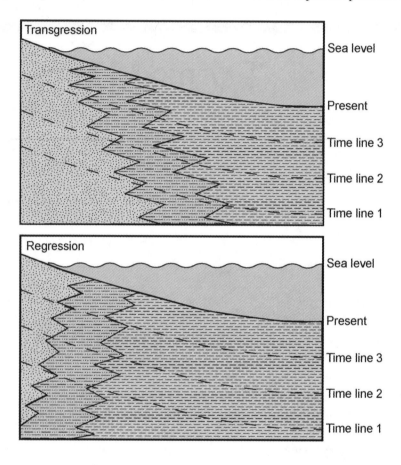

Fig. 7–1. Transgressional (*top*) and regressional (*bottom*) environments indicating changes in sea level, subsidence or uplift rates, or rates of seafloor spreading. Transgressional environments are also described as fining or deepening upward and indicate the shoreline is moving inland. Regressional environments are also described as coarsening or shallowing upward and indicate the shoreline is moving seaward.

Identifying the facies of the particular interval or unit (e.g., lagoonal, marine, terrestrial, and below *storm wave base*) will permit the use of analog environments to model potential exploration trends. Particular facies commonly have more porosity and permeability (fig. 7–2). Possible stratigraphic traps can also be identified when a facies map is used in conjunction with structure contour and isopach maps to locate possible up-dip *pinch outs*, unconformity, or isolated stratigraphic plays (fig. 7–3).

To properly construct and interpret a detailed facies map it is necessary to have a regional understanding of the geology and some outside stratigraphic information (i.e., *core* or *cuttings*); detailed classifications are needed to constrain the probable depositional depth and environment (e.g., *fluvial, alluvial,* marine, or *eolian*). Basic facies maps can be generated based solely on gross lithologies, but for exploration purposes, it is more useful to create a detailed facies map in an attempt to locate potential pay zones or intervals. For example, it is rather simple to identify a limestone in a geophysical log, but unless other stratigraphic information or cuttings are available, it may be impossible to determine if the limestone is *biomicrite* or *intrasparite.* These two types of limestone are deposited in different marine environments and have inherent porosity and permeability differences.

To generate a facies map, lithofacies must be properly interpreted in the geophysical logs and then correlated between wells. A cross section is constructed based on these correlations. It is important to remember that lithofacies correlations are not commonly chronostratigraphic correlations unless there is independent age data (i.e., microfossils, a dated bentonite or non-diachronous unconformity) associated with each lithofacies to corroborate this interpretation. Typically, time lines will cross lithofacies because at a given time different facies are deposited in adjacent deposition settings (fig. 7–1). These facies *interfinger* with one another and migrate back and forth, depending on changes in sea level, the amount of accommodation space, and sediment supply rates. Overall, if sea level rises (a transgression), the amount of accommodation space will increase (as long as sediment supply rates remain constant or decrease) and the facies will deepen or fine upward in a vertical column. The opposite scenario will cause a regression, and the facies will become shallower or coarsen upward. Identifying the direction in which the sediments coarsen will indicate the source region or provenance of the sediment.

Lithology	Porosity	Permeability	Common Depositional Enviroments
Breccia	High	Medium	High energy, fluvial, unconformity, alluvial fan
Sandstone	High	High	High energy, beach, fluvial, eolian, delta, alluvial fan
Siltstone	Medium	Medium	Medium energy, offshore, floodplain, fluvial, eolian
Shale	Low	Low	Low energy, offshore, delta, fluvial
Limestone	High	Medium	High to low energy, shallow marine, warm climate
Dolomite	Medium	Low	Supratidal flat, shallow marine, diagenetic alteration, reefs (In the lab, low temperate dolomite has not been precipitated)
Salt	Low	Low	Low energy, supratidal flat, enclosed basin, hot and dry climate
Gypsum	Low	Low	Low energy, supratidal flat, enclosed basin, hot and dry climate
Anhydrite	Low	Low	Low energy, supratidal flat, enclosed basin, hot and dry climate
Coal	Low	Low	Low energy, coastal swamp, delta

Fig. 7–2. Inferred porosity, permeability, and depositional environments of common sedimentary rock types. Different carbonate and siliclastic rock types cannot be recognized in geophysical logs and require core, drilling chips, or outcrop data to subdivide the gross lithology, constrain the deposition environment (e.g., fluvial, alluvial, marine, or eolian), and calculate porosity and permeability.

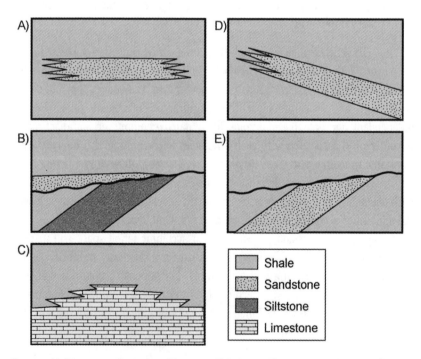

Fig. 7–3. Various types of stratigraphic traps: A) isolated; B) onlap unconformity; C) reef; D) up–dip pinchout; and E) unconformity

EXERCISES

7–1) Identify the lithologies in the following wells (figs. 7–5 through 7–15), and calculate the depth to the top of the bentonite bed. Complete the cross section in figure 7–4 below and draw in a facies map for the unit in which the bentonite bed lies. There is a regional unconformity present in the wells that should be indicated on the cross section.

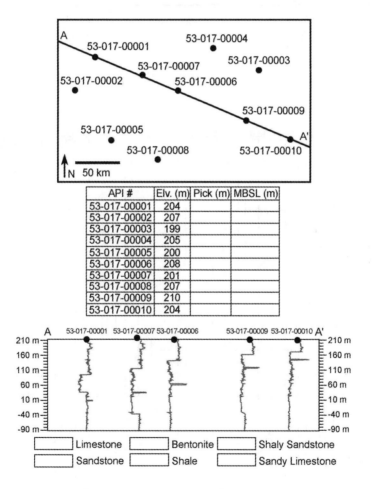

Fig. 7–4

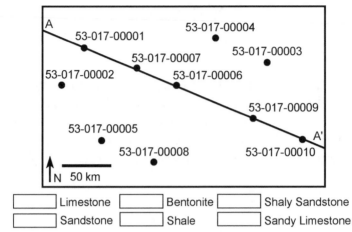

Fig. 7–5

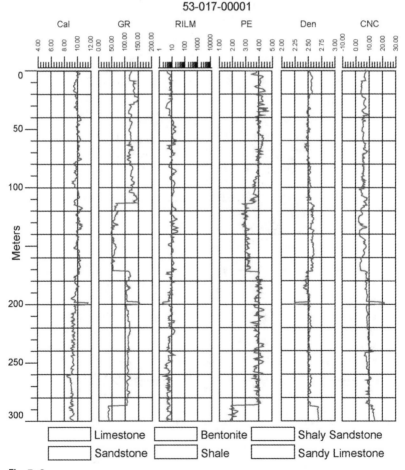

Fig. 7–6

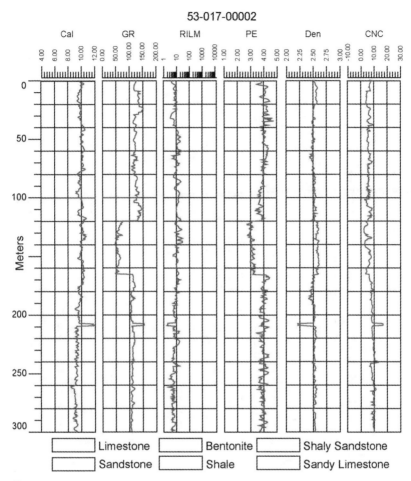

Fig. 7–7

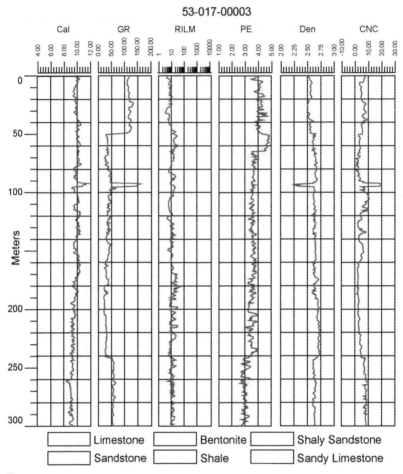

Fig. 7–8

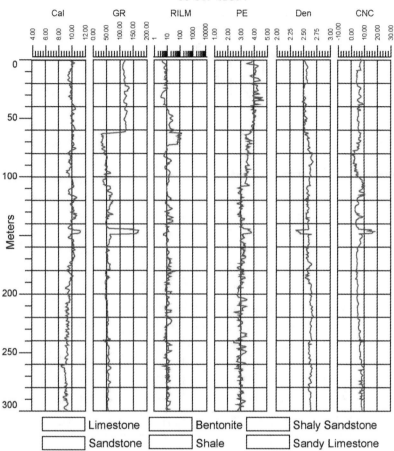

Fig. 7–9

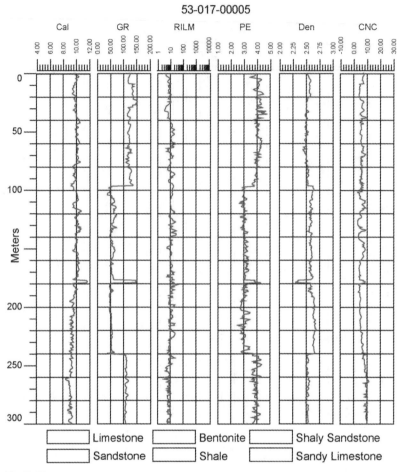

Fig. 7–10

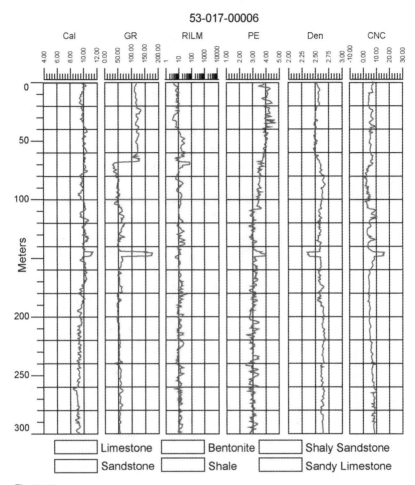

Fig. 7–11

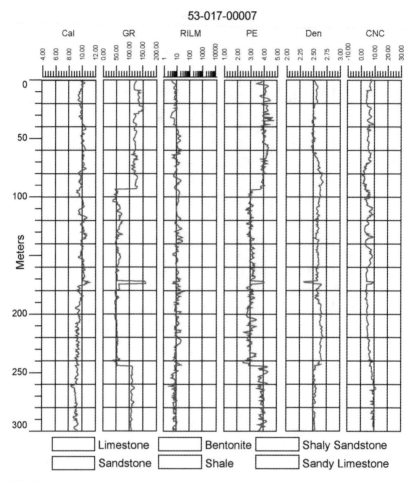

Fig. 7–12

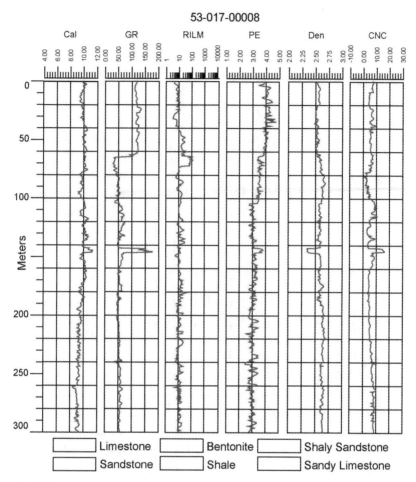

Fig. 7–13

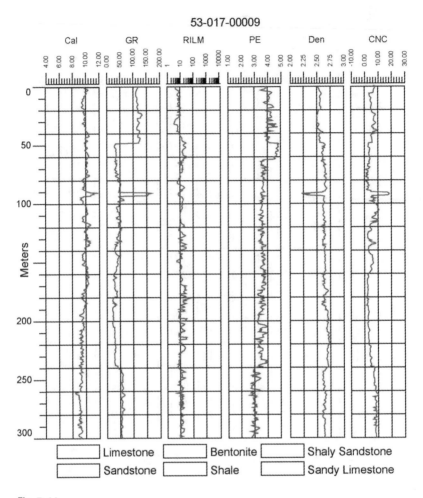

Fig. 7–14

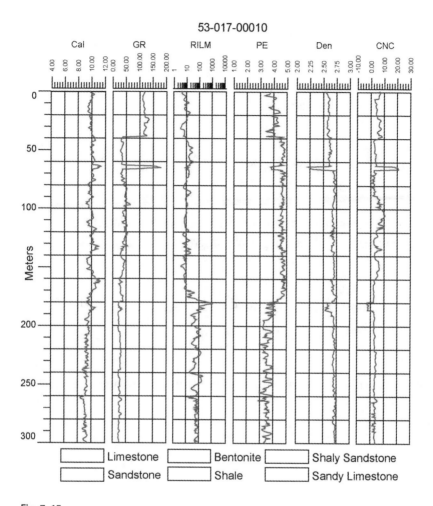

Fig. 7–15

106 Introduction to Well Logs and Subsurface Maps

QUESTIONS

1) Why are facies maps useful?

2) Where is the unconformity located?

3) Based on your cross section, do facies relationships below the unconformity indicate a regional transgression or regression?

4) Based on your facies map, what direction was the paleo-shoreline migrating?

5) Explain how facies maps can help identify potentially large stratigraphic traps.

8

TREND SURFACE
MAPS

Trend surface maps have limited utility in economic exploration, but are very important in understanding regional geologic and stratigraphic changes. Trend surface maps can be used to identify: (1) regional dip (the strike and dip are measured from the trend surface), (2) prominent unconformities (when the strike and dip of stacked maps considerably change orientation; fig. 8–1), (3) the *provenance* for a particular interval or unit, (4) general tectonic changes (i.e., regional uplift or subsidence), and (5) regional deviations (by subtracting the trend surface map from the structure contour map). Using closely spaced wells will only give the local dip and, as a result, it is beneficial to use three widely-spaced data points to better estimate regional dip. This estimation will be different using different data points or wells, so that resulting maps are non-unique.

As with any subsurface map, it is best to make assumptions and interpretations on multiple types of maps representing numerous horizons. Computers can more effectively model trend surfaces using *best fit* algorithms, but an approximate trend surface can be prepared by hand using basic three-point problems. It is important to realize that subsurface maps do not have to correlate with surface (topographic) features and the point data can be derived from any combination of surface and subsurface pick (as long as they are from the same unit).

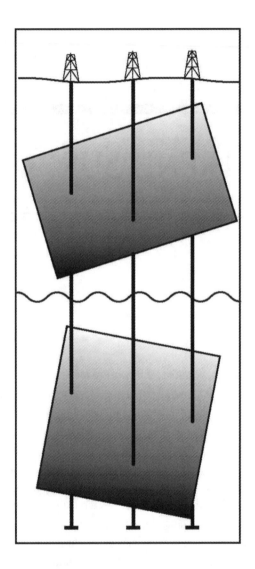

Fig. 8–1. Schematic 3-D view of two trend surface maps separated by an angular unconformity. Light-colored shading represents higher elevations and possible source directions. These surfaces were created from well data only, but any combination of outcrop and subsurface elevation data can be used as long as three known points intersect a given interval or horizon.

THREE-POINT PROBLEMS

It is very difficult to recognize regional stratigraphic changes or structural features from a geophysical well log because of the contrast in scales and the paucity of three-dimensional information. The use of seismic and potential field data will doubtlessly help constrain various stratigraphic and structural features, but these datasets may not be available or may be too expensive for a smaller company. It is therefore important to gain as much information as possible from existing *wildcat* or test wells and surface data. If at least three data points from a correlative horizon (a combination of well and outcrop data) are available, it is possible to calculate the regional dip and generate a basic trend surface map. A trend surface generated from a three-point problem only represents the regional dip based on a limited dataset (three points), and data spacing and structural features can drastically change the estimated dip and direction of the surface. It is more appropriate to use more data or a computer to generate trend surfaces for a more accurate representation of the subsurface.

Solving three-point problems is a basic and important subsurface mapping skill. Three-point problems are a fast way to check regional or local strike and the dip of a unit based on surface or subsurface information. Now that computers can calculate and generate complex surfaces in seconds, it is important to be able to assess whether a computer-generated map is valid. Solving a three-point problem can quickly generate a first-order trend surface and cursorily evaluate the validity of a computer-generated map.

To solve a three-point problem, first reduce the data to subsurface elevations (fig. 8–2a). Find the median subsurface value and locate where that value would be positioned on a line between the maximum and minimum values using a linear gradient (fig. 8–2b). Connect the identical values and record the strike of the line with respect to north (fig. 8–2c). Derive a contour interval appropriate for the scale of the map and draw a line parallel to the strike and space according to the contour interval (fig. 8–2d). The resulting map is a basic trend surface map. In the next chapter we will use these maps to derive even more information about subsurface deviations. A simple method to calculate dip is to use the vertical change between contour intervals and then measure the horizontal distance (perpendicular to strike) between the same two contours using the scale bar (fig. 8–2e).

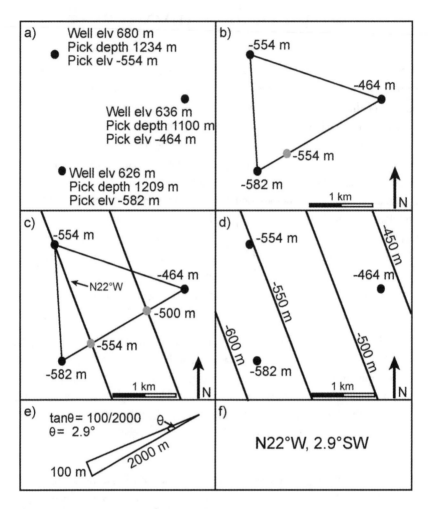

Fig. 8–2. Solving a three-point problem and creating a trend surface map. The data are plotted (a) and a median subsurface value is located between the maximum and minimum values (b). Identical values are connected and the strike is recorded (c). A trend surface is created by drawing spaced parallel lines (d). The dip is then calculated by solving a trigometric problem (e) and the strike and dip values are recorded (f).

Next, calculate the dip using the equation (8.1).

Tan θ = vertical change/horizontal change   (8.1)

The final step is to identify the dip direction with respect to north (fig. 8–2f), which is done by measuring the angle between north and the dip direction. Lastly, record the strike direction and the dip magnitude and direction. For example, a strike and dip of N45°W; 21°NE represents a location where a particular horizon is striking northwest at 45° (315° azimuth) and dipping to the northeast at 21°.

EXERCISES

8–1) Identify the lithologies in the following exploration wells and calculate the depth to the top of the two salt horizons and a bentonite bed. Using figures 8–3 through 8–7, construct three trend surface maps by solving three-point problems.

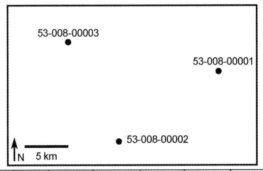

API #	Elv. (m)	Pick1 (m)	MBSL (m)	Pick2 (m)	MBSL (m)	Pick3 (m)	MBSL (m)
53-008-00001	1300						
53-008-00002	1274						
53-008-00003	1271						

Fig. 8–3

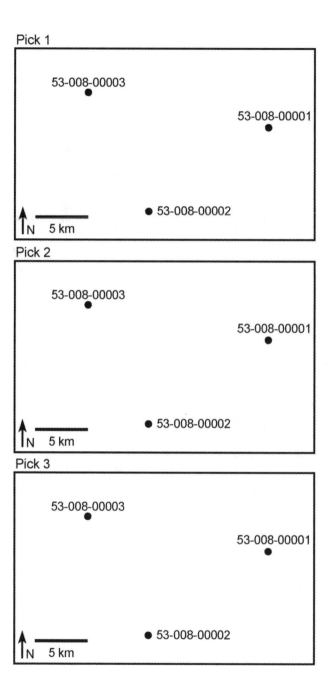

Fig. 8–4

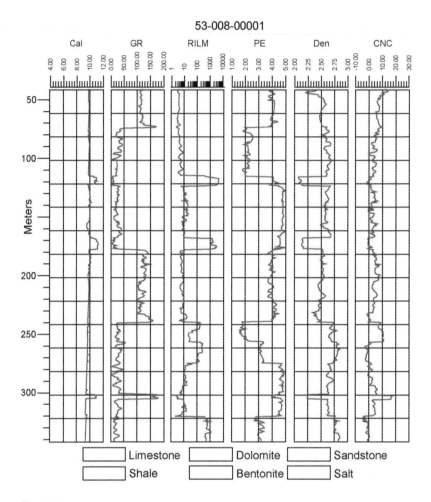

Fig. 8–5

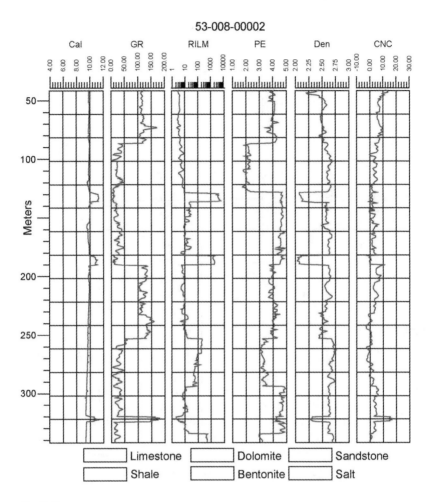

Fig. 8-6

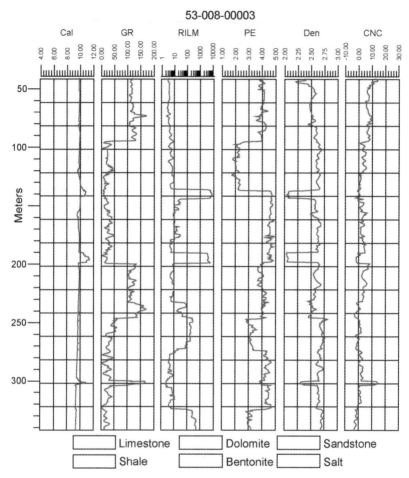

Fig. 8–7

QUESTIONS

1) Why does the regional dip change?

2) What types of interpretation can be made from a trend surface map?

3) What problems may arise if three nearby wells are used to calculate a regional dip?

4) What are the limitations of using a three-point map to estimate regional dip?

5) What is the strike and dip of each horizon?

9

TREND SURFACE RESIDUAL ANOMALY MAPS

Trend surface residual anomaly (TSRA) maps are not widely used because of the increased dependence on 3-D seismic reflection data. Residual anomaly maps are frequently prepared from magnetic and other types of geophysical data, and a similar method for removing the regional trends is used. The process of removing a regional trend from well data is particularly useful in the identification of *blind* (not observable on the surface) structures or structural trends in areas with a large amount of well data. These maps have been used to detect or delineate individual faults, deformation caused by the faults, structural trends, reef structures, and *sole horizons* not identifiable in structure contour or thickness maps (figs. 9–1 and 9–2).

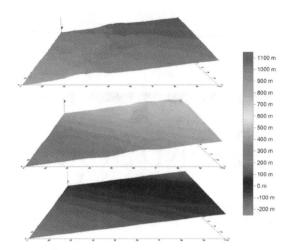

Fig. 9–1. Stacked structure contour maps illustrating a general dip to the lower left. A small-displacement thrust fault is situated near the middle of the upper maps, but its position is masked by the regional dip.

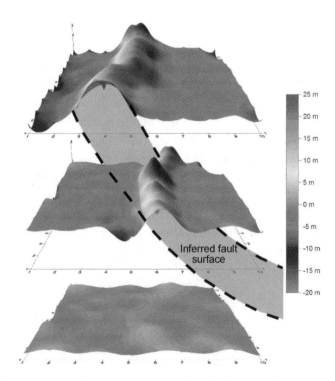

Fig. 9–2. Stacked trend surface residual anomaly maps that correspond with structure contour surfaces. The maps effectively remove the first-order regional dip and magnify deviations from the model. The location of the inferred thrust fault is noticeably easier to identify in the TSRA maps.

MAKING A TSRA MAP

Residual anomaly values can be derived by subtracting modeled trend surfaces from subsurface pick elevations (structure contour data). A trend surface map must first be generated using a regression matrix available in many geological software packages or by estimating a first-order trend surface using a basic three-point problem (fig. 9–3). After a different order trend surface has been generated, the surfaces can be subtracted from observed data. The resulting residual anomaly values are then plotted and contoured. As a general rule, higher-order trend surfaces mimic the observed data too well, and therefore the residual anomaly values are of limited use. First- and second-order residuals are commonly used in areas with small horizontal extent and horizontal stratigraphy or in larger areas with limited data; third-order or higher residuals are usually used in regional mapping (fig. 9–4).

The order of the trend surface is mainly dependent on correlations of known values (e.g., fault displacements or reef thicknesses) from wells within the mapped area. Statistical tests may not be appropriate guides for the selection of degree of the residual anomalies because the targets are ultimately qualitatively chosen based on geologic correlations and not by "goodness of fit" (Davis, 2002). TSRA values below known values limit the effectiveness of the map because observed values are larger. Consequently, a lower-order trend surface residual anomaly map should be generated until the residual anomaly values are slightly above observed or estimated fault displacements or reef thicknesses. For example, a thrust fault identified in a geophysical log with approximately 25 m (82 ft) of offset preferably should reasonably match the TSRA value. Therefore, if a TSRA value derived from a first-order trend surface were 28 m (92 ft), the map would be acceptable. If a second-order TSRA value is calculated to be 23 m (75 ft), the map should be disregarded because the TSRA value is less than the known fault displacement. If possible, multiple observed values should be used to tie together residual values and observed data. This illustrates the subjective and iterative nature of trend surface mapping. However, knowledge of the regional geology can help decrease misinterpretations.

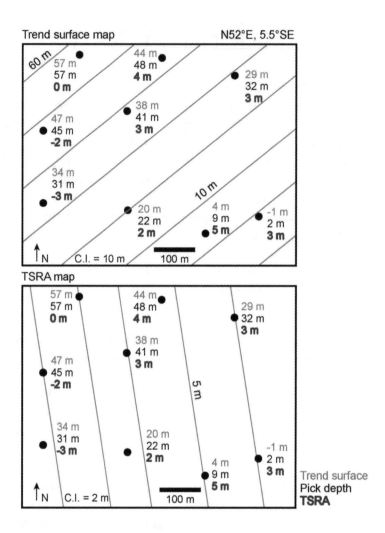

Fig. 9–3. A first-order trend surface map and a corresponding trend surface residual anomaly map generated by subtracting the structure contour data (adjusted pick depth) from the trend surface.

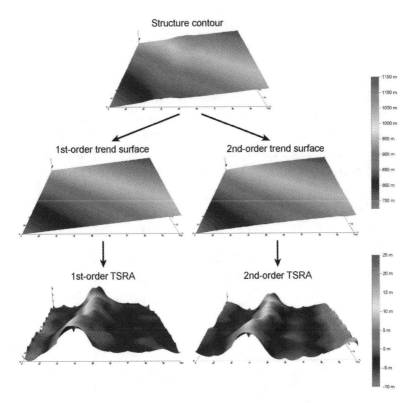

Fig. 9–4. A general TSRA map flowchart. Trend surfaces, which approximate regional dip, are generated from a structure contour map. Trend surfaces are then subtracted from the structure contour map to yield a corresponding TSRA map. Higher-order TSRA maps will have less overall variation because higher-order trend surfaces more closely represent the structure contour map. Known values (e.g., fault displacements or reef thicknesses) should then be cross-checked with the resulting maps to see if any correlations can be made and which order TSRA map better matches the observed or calculated data.

EXERCISES

9–1) Identify the lithologies in the following wells (figs 9–5 through 9–16) and calculate the depth to the top of an anhydrite interval. Solve a three-point problem using wells 53-009-00001, 53-009-00002, and 53-009-00006 and generate an approximate trend surface map. Estimate the depth to the top of the anhydrite (Est.) using the trend surface map and then subtract the trend surface from structure contour data to derive a trend surface residual anomaly value (Res.). Then contour the TSRA values and complete the cross section based on all the information.

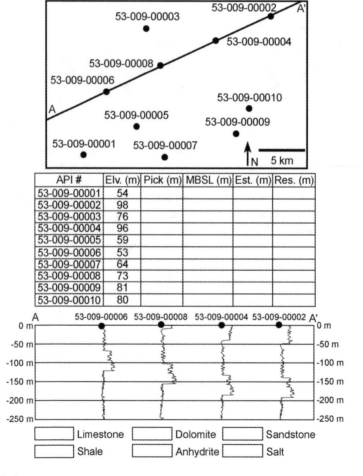

API #	Elv. (m)	Pick (m)	MBSL (m)	Est. (m)	Res. (m)
53-009-00001	54				
53-009-00002	98				
53-009-00003	76				
53-009-00004	96				
53-009-00005	59				
53-009-00006	53				
53-009-00007	64				
53-009-00008	73				
53-009-00009	81				
53-009-00010	80				

Fig. 9–5

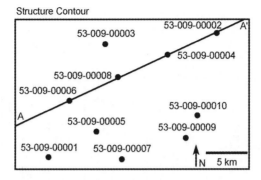

Structure Contour

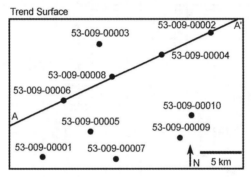

Trend Surface

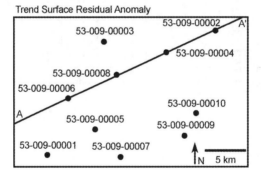

Trend Surface Residual Anomaly

Fig. 9–6

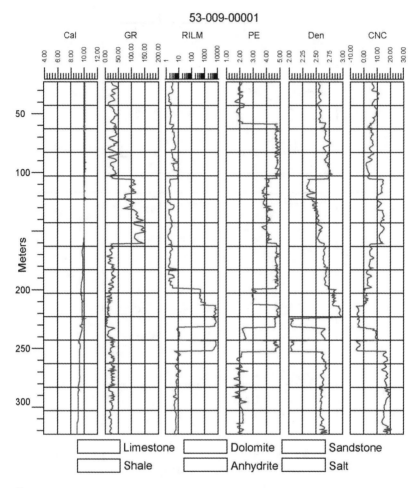

Fig. 9–7

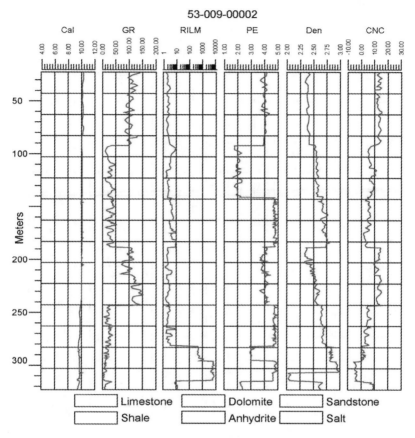

Fig. 9–8

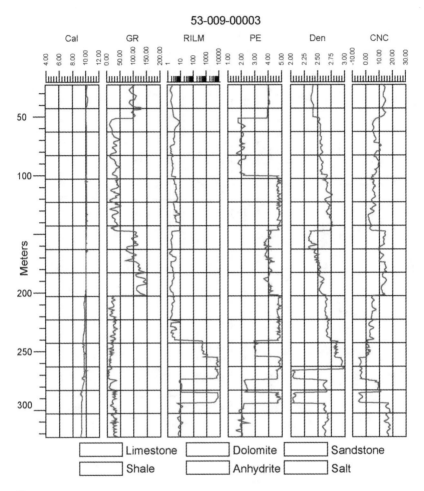

Fig. 9–9

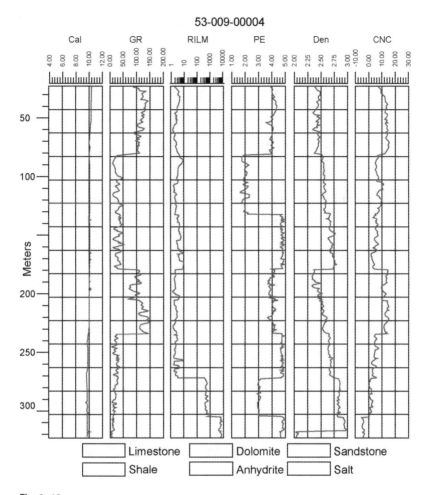

Fig. 9–10

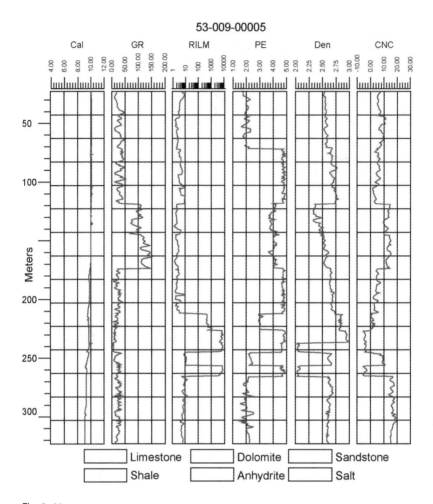

Fig. 9–11

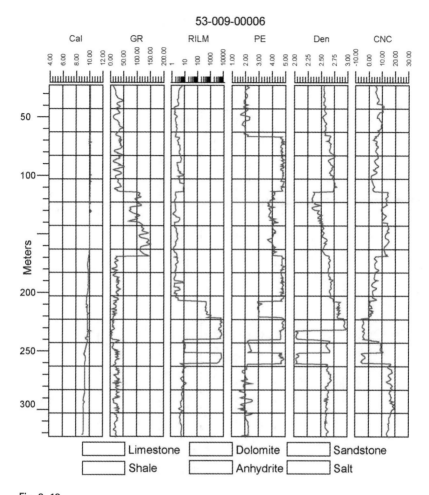

Fig. 9–12

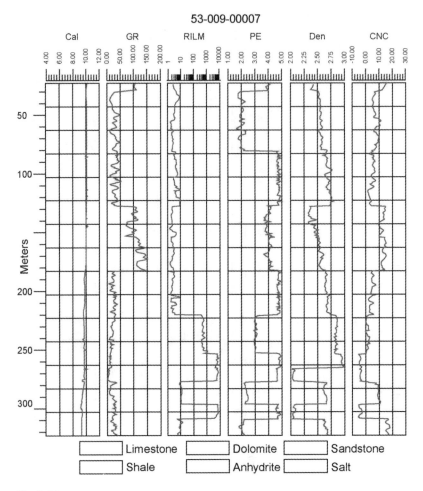

Fig. 9–13

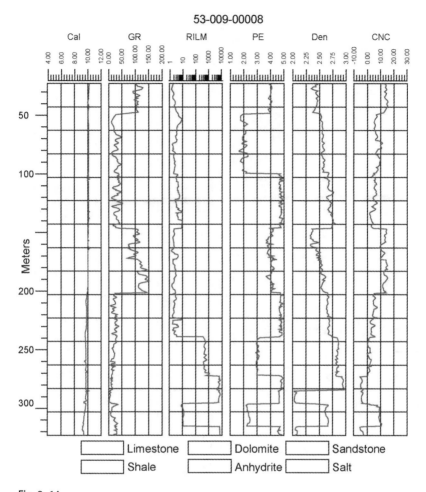

Fig. 9–14

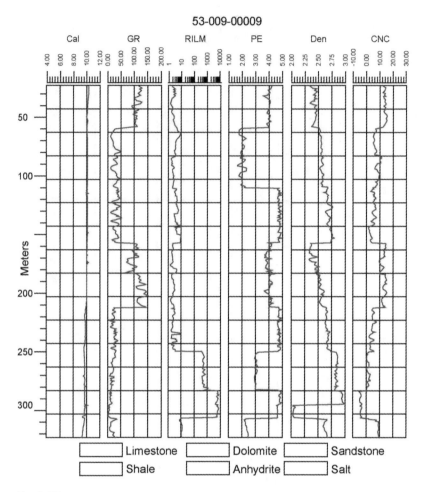

Fig. 9–15

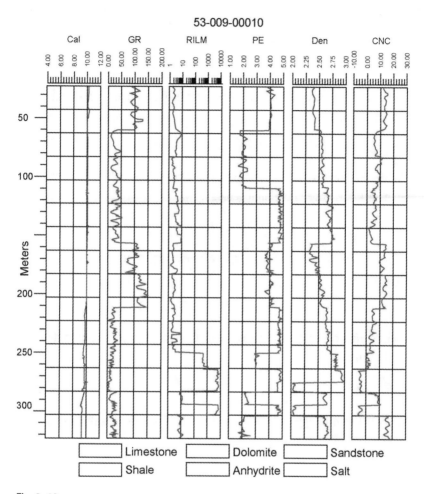

Fig. 9–16

QUESTIONS

1) Why is it important to tie together known displacements to residual anomaly values?

2) When interpreting a TSRA map of a particular area, why is it important to have knowledge of regional geology?

3) What is the advantage of looking at a TSRA and structure contour map together?

4) What types of structures or structural trends have been delineated in TSRA maps?

5) How is the order of the trend surface used to calculate the residual anomaly values chosen?

10

HYDROLOGIC MAPS AND INJECTION WELLS

HYDROLOGIC MAPS

Groundwater flow is often difficult to model. As the amounts of available drinking water and hydrocarbon resources decrease, it is becoming clear that knowing how to construct and interpret different types of hydrologic maps is a highly desirable skill. Potential aquifers are lithologically similar to hydrocarbon reservoirs because they share the same properties (good porosity and permeability). The type of fluids found in the unit and the general depths of the two types of reservoirs are the main differences. Aquifers or reservoirs can vary from sandstone, limestone, and glacial deposit to fractured igneous and metamorphic rock.

Groundwater flow can be estimated by identifying possible aquifers and mapping the depth to the *water table* (fig. 10–1). Since the upper portion of a well is typically cased to protect potential aquifers and to increase the stability of the borehole, it is difficult to identify the water table in these wells. In water wells, the water table is identified while drilling and noted in the *driller's log*. If it was not noted and the well was not cased, the water table can be identified by locating the first negative deflection in the spontaneous potential (SP) and resistivity logs. A temperature log can also help identity the water table because there is commonly a temperature contrast between the air and the groundwater. This effect would be more noticeable on the down-run of a log rather than the up-run.

In an *unconfined* aquifer, the *hydraulic head* is equivalent to the water table. Hydraulic head is a measurement of water pressure at a particular point in the subsurface and it is commonly related to an elevation above a datum (frequently the base of the *piezometer*). A piezometer is a small-gauge, cased well or impermeable tube placed in the ground to measure the level to which water rises in the well.

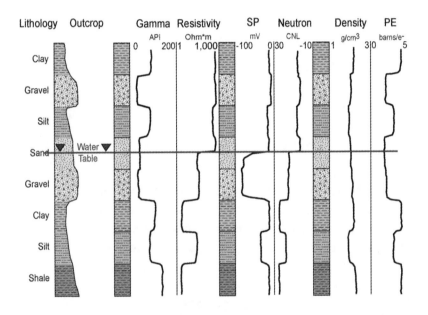

Fig. 10–1. An idealized, uncased well log for saturated and unsaturated strata. Air is highly resistive and has negligible SP because it does not carry an electrical charge. The top section of water wells is always cased to prevent cave-in and aquifer contamination.

Groundwater travels perpendicular to the contoured hydraulic head data (fig. 10–2). A water table map can also be called a piezometric map or *potentiometric surface*. The depth to the water table commonly corresponds to topography, unless the aquifer is *confined*. It is important to understand how groundwater travels in the subsurface when delineating contaminant *plumes*, areas susceptible to contamination (especially in karst areas), and possible locations for water wells or dams. Well logs can also be used to identify contaminants in the subsurface.

Anomalous resistivity spikes are good indicators of certain types of brines and contaminants. In certain cases, it may be advantageous to run other logs (i.e., PE, SP, density, and neutron) to fully delineate a subsurface plume.

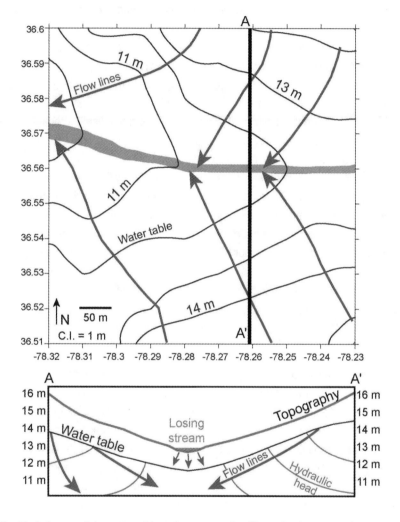

Fig. 10–2. A map of the water table and a cross section illustrating how groundwater travels in the subsurface with respect to hydraulic head

INJECTION WELLS

Groundwater flow is usually thought of as a passive phenomenon. In a pumping or producing well, however, the water around the well is drawn into the borehole and brought to the surface, creating a *cone of depression* around the well. A cone of depression is the conically-shaped area around a well where the water table or potentiometric surface is drawn down. In some hydrocarbon wells, water is used dynamically to *sweep*, or push, hydrocarbons toward producing wells (fig. 10–3). Arrays of producing wells typically surround an injection well. Only two-dimensional, horizontal flow is demonstrated here, although cross flow or vertical flow is more accurate unless there is no vertical connectivity. The injected water not only circularly sweeps oil away from the injection well, it also helps maintain reservoir pressure. The injection well forces water into the subsurface at certain perforated intervals (preferably with potential hydrocarbon reserves) in an attempt to drain more of the reservoir. Injection wells may also use gas or steam to push the hydrocarbons, but water is the most common choice. One reason for using water is that hydrocarbon wells produce waste water, which can be treated and recycled into the injection well, thereby increasing efficiency. Injection wells are most effective and efficient when the injection well is down-dip from the producing wells; it is perforated only in the producing horizons, the produced water is recycled, and multiple producing wells are supported by one injection well.

The amount of time for the water to travel a certain distance is called a *travel time* or *step time*. Higher porosity and permeability intervals have higher travel times (figs. 10–4 and 10–5). When the first injected water is produced, it is commonly called the *breakthrough* point or breakthrough time. The time it takes to sweep an area is dominantly controlled by the permeability of the unit, not the porosity. Porosity is a major factor in estimating *in-place* hydrocarbon reserves, but permeability is the controlling variable in estimating recoverable reserves. Reservoir heterogeneity is a common problem when estimating hydrocarbon reserves and sweep efficiencies. It is common to take a small core sample from each reservoir and estimate the properties of the entire reservoir from the sample. This oversimplification is dealt with by incorporating surface analog data, more core samples, and detailed geophysical modeling.

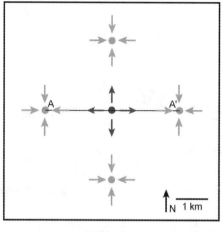

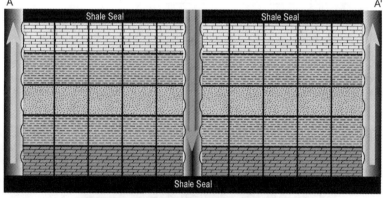

Color	Lithology	Porosity	Permeability	Travel Time
	Sandstone	30%	2000 md	1 d
	Limestone	20%	1000 md	2 d
	Dolomite	2%	50 md	400 d
	Shale	0.1%	1 md	2000 d

Fig. 10–3. Schematic map and cross section views of an injection well surrounded by producing wells. Plausible porosity, permeability, and travel time values were chosen for the same common lithologies (md = millidarcys).

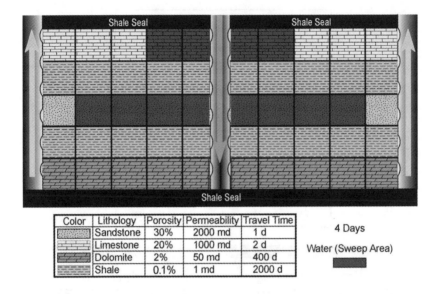

Color	Lithology	Porosity	Permeability	Travel Time
	Sandstone	30%	2000 md	1 d
	Limestone	20%	1000 md	2 d
	Dolomite	2%	50 md	400 d
	Shale	0.1%	1 md	2000 d

4 Days

Water (Sweep Area)

Fig. 10–4. A 2-D view of the swept or sweep area after four days. In this model, breakthrough will occur in the sandstone unit in five days.

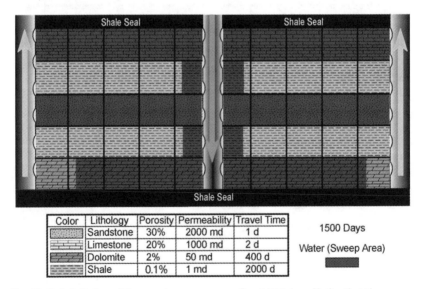

Color	Lithology	Porosity	Permeability	Travel Time
	Sandstone	30%	2000 md	1 d
	Limestone	20%	1000 md	2 d
	Dolomite	2%	50 md	400 d
	Shale	0.1%	1 md	2000 d

1500 Days

Water (Sweep Area)

Fig. 10–5. A 2-D view of the swept or sweep area after 1,500 days. Notice that the dolomite and shale units have not been fully swept and that breakthrough has occurred. In this example, injection would have likely stopped after 10 days because the reservoirs would have been fully swept.

EXERCISES

10–1) Contour the hydraulic head data and draw flow lines. If this were a subsurface hydrocarbon reservoir, where would be the best place to drill a well? Mark an X on the map to locate the well in figure 10–6.

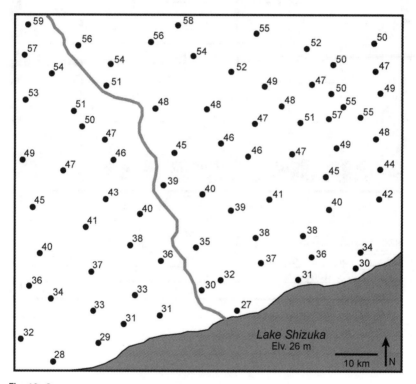

Fig. 10–6

10–2) This complex channel system is a producing hydrocarbon horizon (fig. 10–7). Color each square using the *dominant* lithology's color. Red lines are channel boundaries.

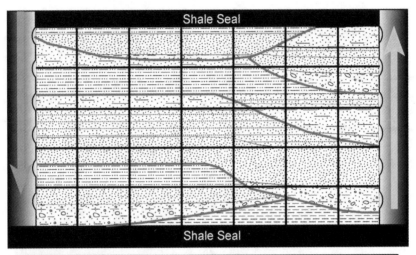

Color	Lithology	Porosity	Permeability	Travel Time
	Coarse-grained sandstone	30%	2000 md	1 d
	Conglomerate	20%	1000 md	2 d
	Fine-grained sandstone	20%	500 md	4 d
	Shaly sandstone	5%	200 md	10 d
	Siltstone	1%	20 md	100 d
	Shale	0.1%	1 md	2000 d

Fig. 10–7

10–3) Using the dominant lithologies from exercise 10–2, shade in the area swept after 10 days (fig. 10–8). Remember to use the travel time of the dominant lithology for the entire box and assume there is no vertical flow.

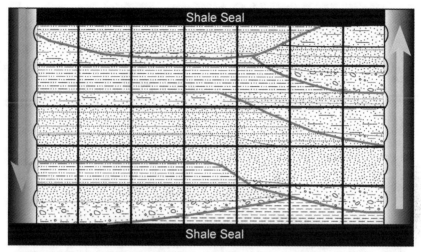

Color	Lithology	Porosity	Permeability	Travel Time
	Coarse-grained sandstone	30%	2000 md	1 d
	Conglomerate	20%	1000 md	2 d
	Fine-grained sandstone	20%	500 md	4 d
	Shaly sandstone	5%	200 md	10 d
	Siltstone	1%	20 md	100 d
	Shale	0.1%	1 md	2000 d

Fig. 10–8

10–4) Using the dominant lithologies from exercise 10–2, shade in the area swept after 100 days. Remember to use the travel time of the dominant lithology for the entire box and assume there is no vertical flow in figure 10–9.

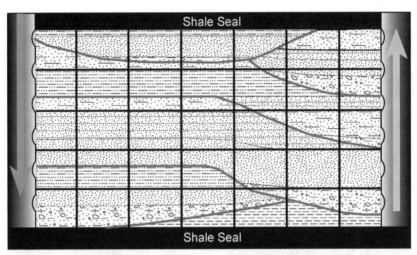

Color	Lithology	Porosity	Permeability	Travel Time
	Coarse-grained sandstone	30%	2000 md	1 d
	Conglomerate	20%	1000 md	2 d
	Fine-grained sandstone	20%	500 md	4 d
	Shaly sandstone	5%	200 md	10 d
	Siltstone	1%	20 md	100 d
	Shale	0.1%	1 md	2000 d

Fig. 10–9

10–5) Using the dominant lithologies from exercise 10–2, shade in the area swept after 500 days in figure 10–10. Remember to use the travel time of the dominant lithology for the entire box and assume there is no vertical flow.

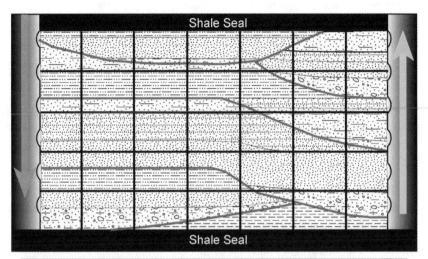

Color	Lithology	Porosity	Permeability	Travel Time
	Coarse-grained sandstone	30%	2000 md	1 d
	Conglomerate	20%	1000 md	2 d
	Fine-grained sandstone	20%	500 md	4 d
	Shaly sandstone	5%	200 md	10 d
	Siltstone	1%	20 md	100 d
	Shale	0.1%	1 md	2000 d

Fig. 10–10

QUESTIONS

1) Why are hydrocarbon wells always cased near the surface?

2) Name two common rock types that are good reservoirs/aquifers.

3) Why is hydraulic head not always equal to the water table?

4) What are the advantages of using injection wells?

5) Why is it not a good idea to inject water into a shale horizon?

11

FORMATION FLUID INTERPRETATION AND HYDROCARBON RESERVES

Identifying the type of fluid in a formation is one of the most important steps in properly developing a field. If the formation fluid type or producing intervals are misinterpreted, a potentially productive well may ultimately be *plugged and abandoned*. The fastest way to identify hydrocarbons in a geophysical log is to look for a porous section of sandstone or limestone that is highly resistive (fig. 11–1). Remember that high resistivity can also be related to the lack of porosity. If hydrocarbons are present, it is important to properly identify the gas/oil and oil/water contacts (fig. 11–2) because they will influence how the field is developed and the estimated amount of recoverable hydrocarbons. The gas/oil contact can be estimated by locating the position in a geophysical log at which the *gas effect* (the crossover effect in neutron and density logs) decreases and the resistivity drops slightly (because gas is more resistive than oil). The spontaneous log can also be used to locate this contact, but it is not as accurate. The oil/water contact can be estimated by noting where the resistivity decreases significantly in a suitable reservoir. These simplifications are useful for the quick identification of hydrocarbons in particular pay zones, but it is also important to know what percentages of hydrocarbons are in the reservoir.

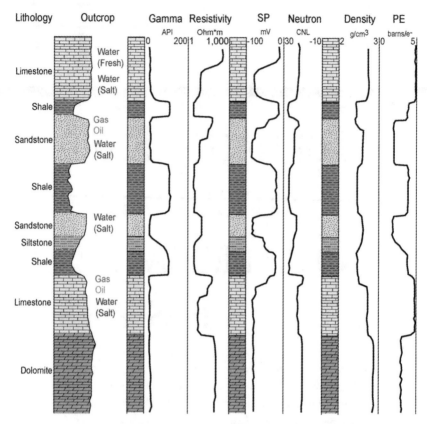

Fig. 11–1. Common geophysical responses to different types of formation fluids. Resistivity, spontaneous potential, neutron, and density logs are the main logs used to identify gas/oil and oil/water contacts; whereas, gamma ray and photoelectric logs are mainly used for lithologic determination.

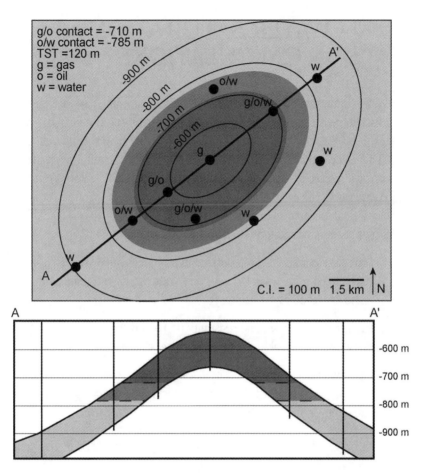

Fig. 11–2. A structure contour map on top of a 120 m (394 ft) thick, hydrocarbon–filled interval. The field is a doubly plunging anticlinal trap with gas/oil and oil/water contacts at –710 m (–2,329 ft) and –785 m (–2,575 ft), respectively.

WATER SATURATION CALCULATIONS

Water saturation (S_w) calculations are a fundamental part of reservoir analysis and reserve calculations. To calculate the water saturation for a particular horizon, it is necessary to first know or calculate the *formation water resistivity* (R_w). This value can be sometimes used for an entire field, but in complex fields or in isolated reservoirs, it may be necessary to use separate R_w values. *Tortuosity* and *cementation* values must also be used to calculate R_w. Tortuosity is the distance a fluid must travel to pass through a given interval. To calculate the R_w and S_w it is necessary to perform the following steps:

1. Look for a porous reservoir horizon in a geophysical log that is probably 100% filled with water or a section with the lowest gamma ray and resistivity readings to calculate R_w.

2. Read the porosity (Φ) and resistivity (R_0) and determine the lithology in this section (it probably will be a sandstone or limestone).

3. Input values for the tortuosity factor (a) and cementation exponent (m) for the identified lithology (sandstone $-$ a $= 0.81$, m $= 2$; limestone $-$ a $= 1$, m $= 2$).

4. Calculate formation water resistivity (R_w) using equation (11.1).

$$R_w = R_0 / (a/\Phi^m) \tag{11.1}$$

5. Look for a section with possible hydrocarbons or a section with low gamma ray readings and the highest resistivity readings to calculate water saturation (S_w).

6. Read the porosity (Φ) and resistivity (R_t) and determine the lithology in this section (it should be either a sandstone or limestone).

7. Input values for tortuosity factor (a) and cementation exponent (m) for the identified lithology (sandstone $-$ a $= 0.81$, m $= 2$; limestone $-$ a $= 1$, m $= 2$).

8. Add the saturation exponent (n) for the identified lithology (both sandstone and limestone have n values of 2).

9. Calculate water saturation using *Archie's equation* (11.2).

$$S_w = ((a \times R_w) / (\Phi^m \times R_t))^{\wedge(1/n)} \tag{11.2}$$

10. Calculate the oil saturation (S_o) using equation (11.3).

$$S_o = 1 - S_w \tag{11.3}$$

For example, if a fully water-saturated sandstone interval had 5% porosity and 0.2 Ω of resistivity, the R_w would be 0.00062. If a potential limestone hydrocarbon reservoir in the same borehole had 15% porosity and 15 Ω of resistivity, the S_w and S_o would be 4.3% and 95.7% respectively.

$R_w = 0.2 / (0.81/0.05^2)$

$R_w = 0.00062$

$S_w = ((1 \times 0.00062) / (.15^2 \times 15))^{\wedge(1/2)}$

$S_w = 0.043$

$S_o = 1 - 0.043$

$S_o = 0.957$

If a regional R_w value is known, only steps 5 through 10 will be required to calculate an interval's oil saturation. Other types of reservoirs (i.e., fractured shale and hydrothermal dolomite) require different equations and are very difficult to model due to the heterogeneity of the reservoir and the larger variance in permeability. Estimating potential hydrocarbon reserves in these reservoir types becomes difficult.

CALCULATING HYDROCARBON RESERVES

Calculating hydrocarbon reserves is an important area of petroleum geology and one that is commonly overlooked. Forecasting reserves that are too high will financially damage a company when production does not meet the estimates. If the calculated reserves are too low, it is possible the field may not be developed at all. There are many types of reserve calculations (e.g., in-place reserves, or total reserves, and recoverable reserves).

This section will cover two basic ways to calculate oil *resources* and reserves that are fast and relatively accurate. Resources are all quantities of useful materials estimated to be in place or recoverable that can be economically extracted at some point in time. Reserves are proven, economically recoverable resources at the current time. A marginal oil field or well may pass below an economical limit and have to be *shut in* (the hydrocarbons remaining *behind pipe* are now a resource).

The first attempt to estimate potential recoverable reserves uses assumed or modeled reservoir properties to estimate a volume of hydrocarbons in a specific area of closure. An area of closure is the region below a given concave down surface. As a general rule, only reservoir rocks with proper *seals* and large areas of closure are prospected. If hydrocarbons in the reservoir have an adequate seal, hydrocarbons will migrate upward and accumulate in an area with closure until the trap is full. At this point, the hydrocarbons will spill out of the trap and continue to migrate upward. The elevation or elevations (if the area of closure is inclined) that denote the base section of the closure is called the *spill point(s)*. Knowing the spill point is essential when initially estimating a target's potential recoverable reserve. Equation (11.4) may yield quick results, but may not be very accurate. It will, however, give an early assessment of the target and indicate whether further exploration or development and production are appropriate. A more accurate volume can be calculated using a *planimeter* to measure the area between certain seismic contours in a given region.

potential resources = [(area × TVT × S_0 × RR × Φ) / FVF] ×
(6.29 bbl or 35.30 ft^3)   (11.4)

where:

area = aerial surface area of the trap or area of closure (m^2)

bbl = barrels

TVT = true vertical thickness, or net to gross (m)

S_0 = oil saturation (%)

RR = recovery rate (%)

Φ = average porosity of the interval (%)

FVF = formation volume factor

The values at the end of the equation represent the average volume of oil and gas at a given pressure ($1 \text{ m}^3 = 6.29$ bbl of oil or 35.30 ft^3 of natural gas). A net to gross, pay, or net thickness may be used in place of TVT if the reservoir is heterogeneous and only a portion of the interval is a suitable reservoir. The *formation volume factor* (FVF) is a variable that takes into account the amount of volume loss due to dissolved gas coming out of solution during extraction. The value of the factor is commonly between 1 and 1.5.

For example, if a 50% oil-saturated sandstone interval is 10 m (33 ft) thick, has 15% porosity, is located in a field (area of closure) covering 12 km^2 (4.6 mi.^2), and has an average recovery rate of 25%, the potential oil resources in the interval (FVF = 1.1) will be approximately 12,866,000 bbl.

area = $12,000,000 \text{ m}^2$ ($14,351,880 \text{ yd}^2$)

TVT = 10 m (33 ft)

RR = 25%

Φ = 15%

S_0 = 50 %

FVF = 1.1

A second method to estimate potential reserves and economical reserves is to use production data and well *decline curves* to project ultimate yield of a well(s) with the assumption they are draining the entire trap. This method is more accurate, but it requires test and production wells along with production data. The method also assumes that the production of the well(s) will continue to decrease along the modeled decline curve (fig. 11–3), which is commonly plotted on a linear graph. To calculate the recoverable reserves, it is necessary to graph current production data and extrapolate the data. The area beneath the modeled curve is equal to the total amount of recoverable reserves. There are two ways to calculate this value; 1) calculate the area under the curve using the equation of the decline curve; or 2) plot the data on a logarithmic scale and calculate the yearly and cumulative production based on the average yearly production. Next, add each year's production to find the total recoverable reserves.

The latter method is less accurate, but doesn't require knowing the equation of the modeled decline curve. In the example below, the average production during the first year was 953 bbl/d, or 347,845 bbl for the entire year.

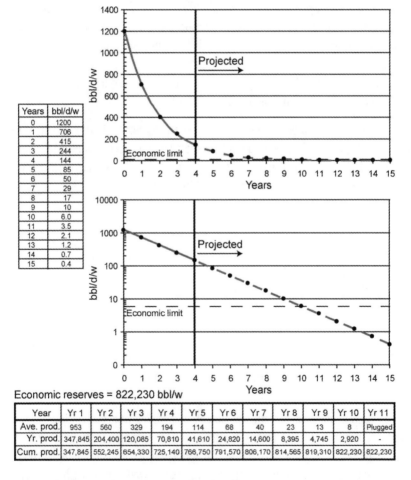

Years	bbl/d/w
0	1200
1	706
2	415
3	244
4	144
5	85
6	50
7	29
8	17
9	10
10	6.0
11	3.5
12	2.1
13	1.2
14	0.7
15	0.4

Economic reserves = 822,230 bbl/w

Year	Yr 1	Yr 2	Yr 3	Yr 4	Yr 5	Yr 6	Yr 7	Yr 8	Yr 9	Yr 10	Yr 11
Ave. prod.	953	560	329	194	114	68	40	23	13	8	Plugged
Yr. prod.	347,845	204,400	120,085	70,810	41,610	24,820	14,600	8,395	4,745	2,920	-
Cum. prod.	347,845	552,245	654,330	725,140	766,750	791,570	806,170	814,565	819,310	822,230	822,230

Fig. 11–3. Two graphs of modeled production data (4 years observed and 11 years estimate) of a well that has an economic limit of 6 bbl. Estimated yearly production and cumulative production are also shown for the 10 years the well will remain economical.

Graphing production data also enables an estimation of when a well or field will fall below the *economic limit*, or cutoff line. The economic limit depends on numerous variables, but is dominantly controlled by market prices. Once a well or field drops below the limit, the well is shut in or abandoned and the remaining recoverable reserves are left in the subsurface. These reserves may be recovered in the future if the economic limit is lowered or if the well could be made to produce more hydrocarbons. Knowing the production data and economic limit makes it possible to estimate the economical recoverable reserves. Economical recoverable reserves can be estimated by calculating the area below the decline curve between initial production and the time when the decline curve crosses the economic limit. In figure 11–3, the economical limit for each well is 6 bbl/d. The wells will, therefore, be shut in after 10 years, having produced 822,230 bbl of hydrocarbon; the remaining reserves will remain in the ground or behind pipe.

EXERCISES

11–1) Identify the lithologies in the well in figure 11–4 (the shaded sections are for water-saturation calculations).

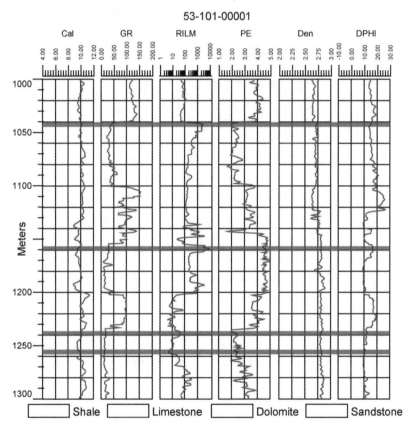

Fig. 11–4

11–2) Draw the decline curve for a theoretical well using the production data, and calculate the economic reserves in figure 11–5.

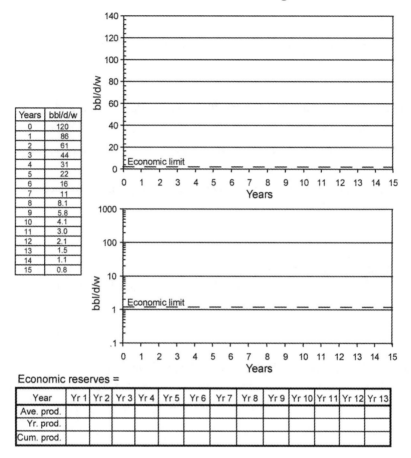

Years	bbl/d/w
0	120
1	86
2	61
3	44
4	31
5	22
6	16
7	11
8	8.1
9	5.8
10	4.1
11	3.0
12	2.1
13	1.5
14	1.1
15	0.8

Economic reserves =

Year	Yr 1	Yr 2	Yr 3	Yr 4	Yr 5	Yr 6	Yr 7	Yr 8	Yr 9	Yr 10	Yr 11	Yr 12	Yr 13
Ave. prod.													
Yr. prod.													
Cum. prod.													

Fig. 11–5

QUESTIONS

1) What type of reservoir must be used to calculate R_w?

2) What does moving the economic limit do to production time and estimated reserves?

3) Why are reserves more accurately predicted using limited production data rather than calculated reservoir properties?

4) Calculate the S_o in the shaded sections (1,041 m, 1,159 m, 1,238 m, and 1,257 m) in well 53-101-00001 (calculate an R_w value first). Show your work.

5) What are the potential reserves (in bbl) of this small horizon? (TST = 2 m, Φ_a = 16°, S_w = 23%, Φ = 26%, aerial surface area = 6,000 m², recovery rate = 20%, FVF = 1.2) (1 m³ = 6.29 bbl of oil)

12

MINING MAPS

The mining industry also uses subsurface maps, along with other types of information, to locate economical mineral deposits. Basic exploration tests involve a systematic drilling program to estimate potential *ore grade* and volume. These variables are then used to evaluate the overall reserves and potential profit. This chapter will not deal with volumes or surfaces, only ore grade.

In the mining industry, porosity is not desirable because void space means a certain volume within a deposit is empty space (i.e., not a recoverable resource). A 3-D map is generated by extrapolating borehole and test data in a similar manner as net pay was calculated in chapter 6. The mining industry does not normally run 3-D seismic surveys or geophysically log test wells because drill core data and descriptions provide more important information, and because economic deposits frequently found in mountainous regions make seismic surveys difficult. It is more common that mining companies run potential field or remote sensing surveys over an area to delineate a deposit's maximum extent and then use test wells and ore grade to further constrain the boundaries of the reserve.

Mineral resource deposits are similar to hydrocarbon plays; the deposits can be broken down into a mineral resource system (they have a source, a reservoir, and a seal). The extent of mineralization, however, differs from hydrocarbon plays. In a hydrocarbon play, the horizontal extent of the play is defined by an area of closure. The horizontal extent of a mineral deposit in sedimentary *host* rocks is dominantly controlled by bedding orientation and attitude, fracture or fault intersections, and stratigraphic changes (fig. 12–1). A host rock is the unit in which the mineral commodity is situated.

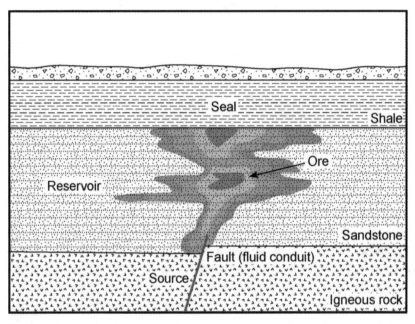

Fig. 12–1. A mineral resource system (source, reservoir, and seal) for a fault-controlled ore deposit. Darker shades indicate higher grades.

CALCULATING MINING RESOURCES AND RESERVES

To create a resource map, the only data required are the location of pick in the subsurface and mineral grade. Potential resource and reserve estimates are calculated in a similar manner as hydrocarbons reserves, except there is a 3D component related to ore grade changes. To calculate the potential resources, use equation (12.1).

$$\text{potential resources} = (\text{area} \times \text{TVT} \times \text{OG}_1) + (\text{area} \times \text{TVT} \times \text{OG}_2)\dots$$
$$+ (\text{area} \times \text{TVT} \times \text{OG}_n) \times \text{AWR} \times \text{density} \tag{12.1}$$

where:

area = aerial surface area of a particular ore grade (m^2)

TVT = true vertical thickness of a particular ore grade (m)

OG = ore grade (%)

AWR = atomic weight ratio (%)

density = density of the ore

To calculate the entire potential resource, it is necessary to calculate the volume reserves for each grade from zero to the maximum grade denoted on the map. This volume is then multiplied by the atomic weight ratio of the mineral, element, or rock. The *atomic weight ratio* (AWR) is a factor that specifically indicates how much of the commodity exists in the ore. If the commodity is an element such as gold or sulfur, the ratio would be a one-to-one ratio, or 100%; if the commodity is a mineral (or rock) such as hematite (Fe_2O_3), the ratio of Fe versus O would be roughly 3 Fe:1 O. Fe and O have atomic masses of 55.84 g/mole and 16 g/mole, respectively. The atomic weight ratio for hematite (Fe:O) is 111.68:159.68, or approximately 70%. This means that a sample of hematite consists of 70% Fe and 30% O. Since most mineral commodities are marketed by weight, both reserve and resource estimates need to be multiplied by the ore's density. In the previous example, hematite's ore is Fe, which has a density of 7.86 g/cm^3, or 7,860 kg/m^3.

To calculate the reserve volume of the deposit, use equation (12.2) to calculate the volume of the ore grades above the economical limit (similar to the difference between hydrocarbon resources and reserves).

$$\text{potential reserves} = (\text{area} \times \text{TVT} \times \text{OG}_1) + (\text{area} \times \text{TVT} \times \text{OG}_2) \ldots$$
$$+ (\text{area} \times (\text{TVT} \times \text{OG}_n) \times \text{AWR} \times \text{density} \tag{12.2}$$

where:

area = aerial surface area of a particular ore grade above an economical limit (m^2)

TVT = true vertical thickness of a particular ore grade (m)

OG = ore grade above an economical limit (%)

AWR = atomic weight ratio (%)

density = density of the ore

EXERCISE

12–1) A new potential zinc (sphalerite or ZnS) mining district was prospected in a remote area. One well was cored and logged, whereas the remaining wells were unlogged. Test wells indicated that the extensively fractured strata is flat-lying and that ore is located in a 10-m thick section in the upper part of a carbonate unit (fig. 12–2). Identify the lithologies in the well and contour the ore grades using a 3% interval (fig. 12–3). Assume there is no topography in the cross section.

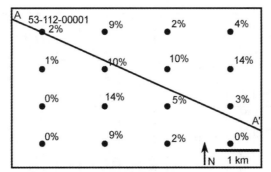

API #	Elv. (m)	Pick (m)	MBSL (m)	Pick2 (m)	MBSL (m)	Pick3 (m)	MBSL (m)	Pick4 (m)	MBSL (m)
53-112-00001	67								

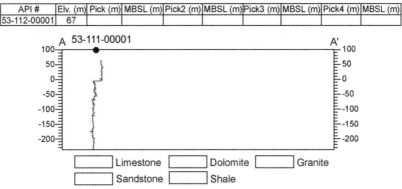

Fig. 12–2

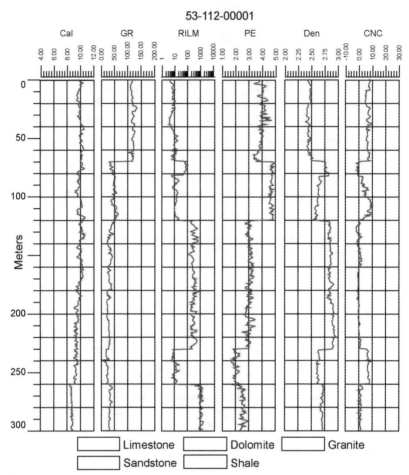

Fig. 12-3

QUESTIONS

1) What is the orientation of the mineralization, and why is the mineralization localized in the upper part of the carbonate unit?

2) If the economical cutoff is 6%, what is the total volume of economical rock in m^3 (assuming it could all be mined)?

3) Calculate the total volume of economically recoverable zinc within the volume calculated in question 2 (atomic weight ratio of sphalerite is roughly 2 Zn:1 S).

4) What is the weight of economically recoverable zinc? The density of zinc is 7,140 kg/m^3.

5) What is the total value of the economically recoverable zinc within the mapped area? Assume the zinc market value is $1/lb, or $2,205/metric tonne (1 metric tonne equals 1,000 kg).

13

CROSS SECTIONS

CROSS SECTION TYPES

In previous chapters and exercises, various profiles and cross sections were illustrated and constructed. One of the most important ways to visualize subsurface features is to create a valid cross section. It is impossible to entirely grasp subsurface stratigraphic and structural architectures, even with the use of 3-D seismic data. It is, however, possible to create a comprehensive 2-D cross section using known or projected data. One can think of the cross section as the portrayal of subsurface outcrop, with sections of good exposure (well-constrained) and poor exposure (poorly constrained). In areas of poor or incomplete exposure, it is necessary to project contacts and structural trends into areas of better exposure using nearby data or information. Commonly, nearby well or seismic data and projected contacts (from contour maps) are used to correlate poorly constrained sections. The danger in using projected data is that it is dependent on various factors (i.e., seismic migration, gridding algorithms, and map scales). For example, two contour maps derived from the same dataset may look significantly dissimilar if they were generated using different gridding algorithms or interpreters.

Cross sections can be divided into two categories, projected and anchored (fig. 13–1). *Projected cross sections* are profiles with no direct well control. This category can be further subdivided into *synthetic* and *bounded cross sections*. Synthetic cross sections are profiles with no direct data to constrain the position of contacts and features; whereas, bounded cross sections utilize nearby data to semi-constrain them (fig. 13–2).

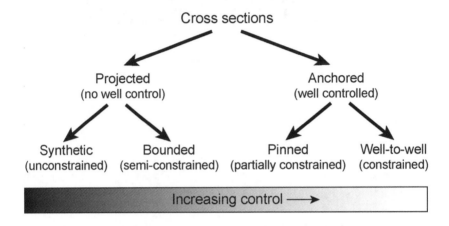

Fig. 13–1. Chart of cross section categories. Cross sections are subdivided by the amount of well and data control.

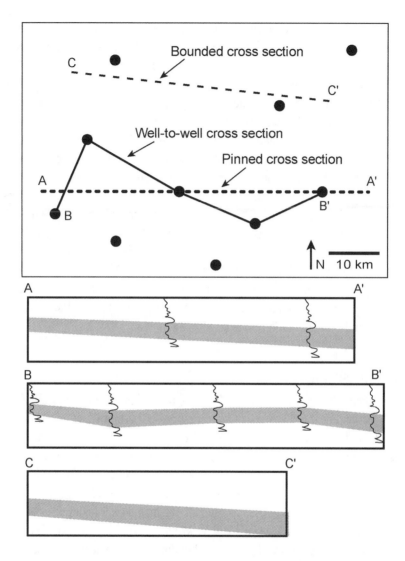

Fig. 13–2. Different cross section types. A marker unit (gray) generally thickens to the east and north. Note the apparent thickening and thinning in the well-to-well cross section caused by the orientation of the section line.

Anchored cross sections are profiles that intersect one or more wells and, therefore, have some degree of direct well control. Well data are anchored, or *tied*, to the cross section because its spatial position is known. It is standard practice to place at least one well log trace (commonly gamma ray, sonic, resistivity, or spontaneous potential) on an anchored cross section. Anchored cross sections can be further subdivided into *pinned* and *well-to-well* cross sections. Pinned cross sections are a profile that intersects one or more wells. Typically, this type of cross section is linear and partially constructed from projected data. Well-to-well cross sections are profiles that correlate multiple wells. These profiles are completely controlled by well data and are commonly non-linear. The main disadvantages of using well-to-well cross sections are they potentially can enhance out-of-plane features and create misleading subsurface relationships. For example, in figure 13–2, the marker unit in B-B' appears to thicken and thin, but this is an apparent change due to the varying section orientation. The marker unit (gray) generally thickens to the north and east as indicated in the other cross sections. Commonly, the best cross section orientation is perpendicular to a desired target, but frequently the orientation of the target is unknown.

Contour slicing

Contour slicing is a manual or electronic method for creating a profile for a given section line (fig. 13–3). Contour slicing can be performed electronically by creating a *blanking file* or section line, and then having a computer routine such as a macro calculate or return the value corresponding to each point where the section line intersects the contoured data. The number of values returned, along with the values themselves, will be a function of the number of nodes in the map and the gridding algorithm used to generate the map. Fewer nodes or generated data points in a mapped surface will yield fewer intersections and, therefore, fewer values. The gridding algorithm will mainly control how well the data are honored in a map (e.g., kriging algorithm routines honor data more closely than minimum curvature algorithms, which is undesirable for representing larger, clustered datasets).

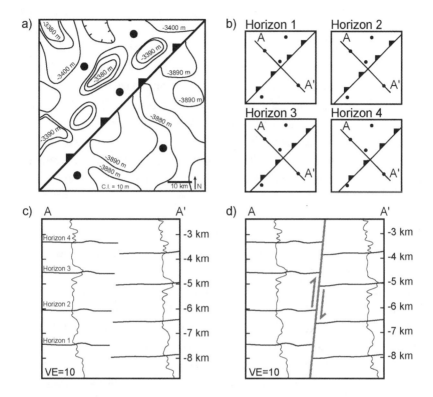

Fig. 13–3. To complete a detailed cross section along a given profile, it is first necessary to standardize the given dataset and produce (a) a series of appropriately spaced, marker horizons (b). After choosing an appropriate section line, each map is contour-sliced and the resulting profiles stacked in a single cross section (c). The profile is then completed by overlaying well data (if available), mapped or inferred structures (d), noting the vertical exaggeration.

Contour slicing by hand is much easier, but not as fast or accurate. Hand contouring takes the place of a gridding algorithm; the transfer of contacts, faults, and well locations to a cross section is performed using graphical software or by paper and pencil. After the values have been transposed to the cross section, the values or dots are connected by a smoothed line that may be broken by faults (if they intersect the line). Usually, the data near a fault break are not as good as those of the surrounding area. This is because a fault break may cause edge effects or contouring artifacts. This problem can be minimized when finishing the cross section by utilizing basic geologic principles, uniform thicknesses, and appropriate fault displacements to stylistically connect the two sides (fig. 13–3d).

Vertical exaggeration

Vertical exaggeration (VE) is a dimensionless number that compares the horizontal and vertical scales of a cross section. It is derived by dividing the horizontal scale by the vertical scale. For example, geologic maps frequently have a 1:24,000 (or 1/24,000) horizontal scale. If these maps have cross sections with a vertical scale of 1:6,000 (or 1/6,000), the vertical exaggeration of these profiles will be 4 (assuming the cross section has the same horizontal scale as the geologic map). The value is dimensionless because the units cancel out. Vertical exaggeration is important because many large-scale or regional studies have cross sections that would be too small (vertically) to effectively depict small-scale features. The drawback of vertically stretching a cross section is that it distorts the shape and size of a feature (fig. 13–4), which may lead to significant misinterpretations or incorrect economic estimates.

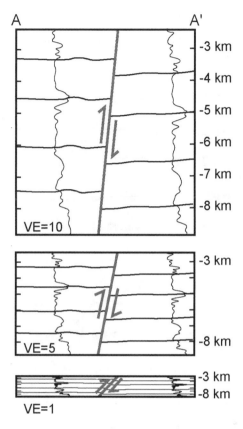

Fig. 13–4. An example of a cross section (same as fig. 13–3d) with different vertical exaggerations. Notice how the thrust fault appears almost vertical in the top example, but at about 45° in the bottom example, which does not have any vertical exaggeration.

Balanced cross sections

Cross sections, just like contouring, naturally look different because interpreters may have different datasets and their own stylistic preferences. Regardless of stylistic preference, it is difficult to determine that a cross section is valid, and a test is required to determine whether a cross section is acceptable. The most common validity tests are called *area, line,* and *volume balancing*, which are easily judged when displayed as a *retrodeformed* or *palinspastically* restored cross section (fig. 13–5). Retrodeformed or palinspastic cross sections are sections that have been restored to their original, undeformed state by some combination of putting together or pulling apart fault blocks and strata. If a deformed section is balanced, the undeformed cross section will fit together with no gaps or overlap. This indicates the area (area balanced) and/or fault and stratigraphic contacts (line balanced) were preserved in the deformed section; therefore, it is a viable solution. Volume balancing is area balancing in 3-D, but is the hardest type of balancing and seldom undertaken.

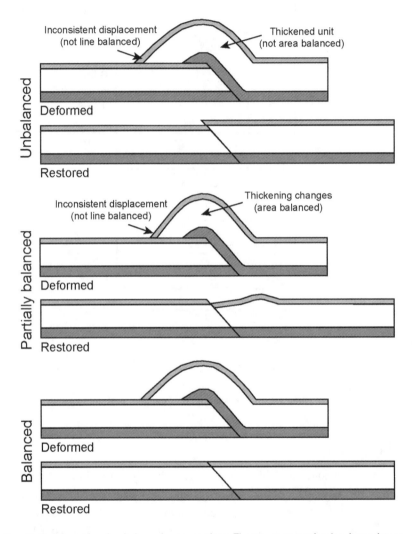

Fig. 13–5. Balanced and unbalanced cross sections. The top cross section has inconsistent fault displacement and the white interval is thicker on the right side of the section. The restored section is, therefore, not balanced (line or area). The middle deformed section has a thickened white interval in the fold hinge and thinned left limb. The section area balances, but does not line balance because the fault displacement is inconsistent. The bottom deformed section area and line balances, and is an invalid model.

There are exceptions and assumptions when judging whether a cross section is valid or plausible. A cross section that balances does not signify that it is the correct solution to the given dataset. This is because in order to create a balanced 2-D cross section, it is necessary to assume the deformed units were (1) originally horizontal, (2) of a constant thickness, (3) not affected by syndeformational sedimentation or out-of-plane mass movement, and (4) all created during one orogenic event.

It is difficult and sometimes impossible to balance sections (in 2- or 3-D) that contain salt domes, reef complexes, an impact structure(s), facies changes, post-depositional dolomitization, and polydeformed strata. This is because each of these features has units with varying thicknesses due to volume loss or gain, and irregular deposition patterns. For example, a pinnacle reef, in a transgressive system, may build upward at the same rate as sea level rises. If this reef is preserved, the lower section of the reef will have a similar facies as the surrounding strata, but the upper portion will be surrounded by a deeper water facies. Lithofacies-based correlations and cross sections across these features will, therefore, not balance because the units will not have constant thicknesses.

In some depositional regimes (e.g., offshore deltas) it is also not possible to balance a cross section because of syndeformational sedimentation. In these areas, the delta is deforming (usually by normal faulting) because of the immense weight of the un- to semi-consolidated sediment. As movement occurs along these normal faults, sediment is contemporaneously deposited over the fault (these faults are called *growth faults*). Sediment concentrates on the down-dropping *hanging wall* because it has more accommodation space than the *footwall*. This ultimately creates a situation where the sediment in the hanging wall is thicker than the footwall. The thickened sediment is commonly called *growth strata* because it is associated with growth faults. Cross sections with growth strata and growth faults are, therefore, unbalanceable.

EXERCISES

13–1) Standardize the given offshore well data to sea level (SL) and calculate the pick elevations. The drilling floor (DF) is 15 m above sea level (SF = sea floor).

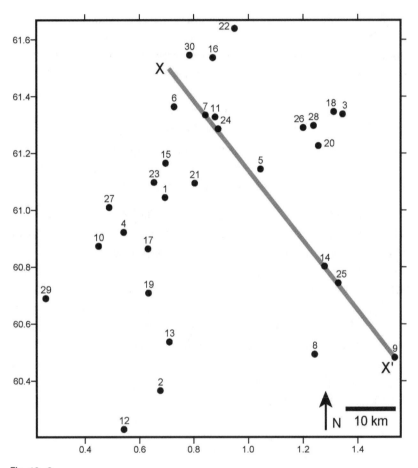

Fig. 13–6.

Well	Datum	SF	Pick A	A Elv	Pick B	B Elv	Pick C	C Elv	Pick D	D Elv
1	DF	-104	190		719		1167		1453	
2	SL	-85	545		1170		1806		2438	
3	DF	-88	441		1270		1660		2109	
4	SL	-79	164		653		1172		1620	
5	DF	-102	471		1327		1694		1836	
6	SF	-86	23		521		867		1020	
7	DF	-100	195		744		1105		1255	
8	DF	-122	575		1642		2142		2703	
9	SL	-105	774		1893		2429		2822	
10	SL	-81	137		627		1158		1701	
11	SL	-118	207		771		1130		1281	
12	SL	-110	498		1171		1829		2506	
13	SF	-108	416		1446		1991		1009	
14	SL	-113	467		1588		2065		4198	
15	SL	-122	109		588		1008		1148	
16	SF	-82	10		602		935		1109	
17	DF	-74	284		845		1380		2039	
18	SL	-83	462		1295		1653		1830	
19	SL	-106	344		998		1607		3410	
20	SF	-84	469		1441		1901		3464	
21	SL	-97	202		766		1161		1287	
22	SL	-95	110		707		1068		1257	
23	SL	-74	106		588		1012		1136	
24	SF	-83	128		681		1072		1218	
25	SL	-80	541		1216		1794		2500	
26	SF	-98	336		1127		1500		1851	
27	SL	-92	86		535		1046		1463	
28	SL	-93	469		1214		1604		1926	
29	SF	-84	34		513		12525		2474	
30	DF	-76	82		612		948		1121	

13–2) Create four structure contour maps using the standardized data from exercise 13–1. Two normal faults have been identified in a previous survey and their mapped positions are shown.

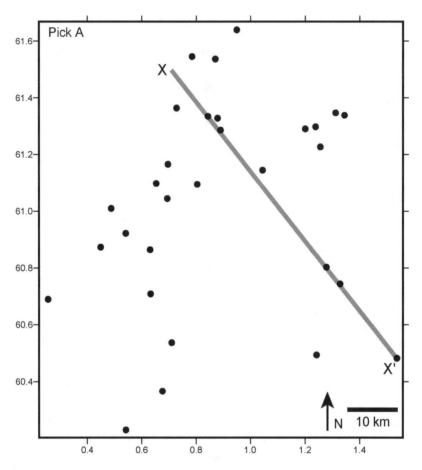

Fig. 13–7.

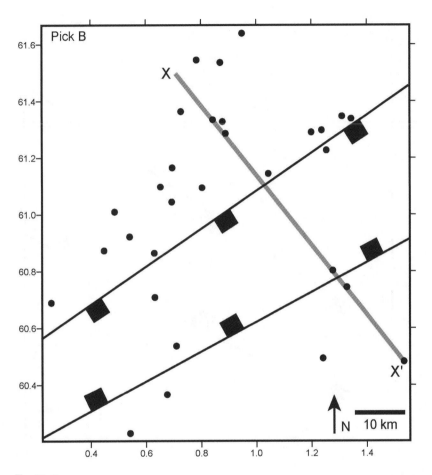

Fig. 13–8.

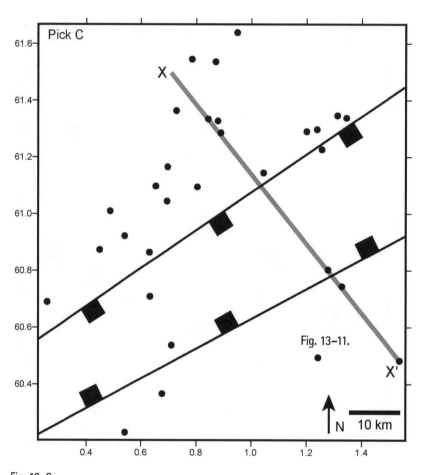

Fig. 13–9.

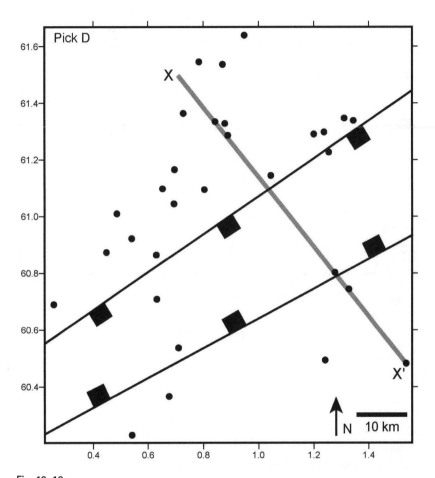

Fig. 13–10.

13–3) Create four profiles (X-X') and stack them in the provided cross section. Complete the pinned cross section. Note: this cross section will not balance.

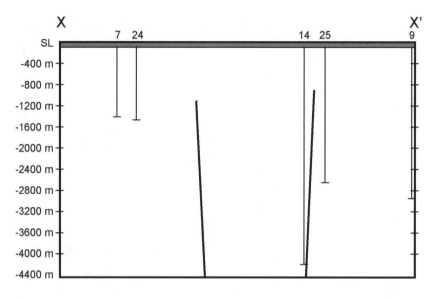

Fig. 13–11.

QUESTIONS

1) What is the difference between synthetic and bounded cross sections?

2) What is a possible significant problem of using a well-to-well cross section?

3) What is the vertical exaggeration of cross section X-X'?

4) Why is cross section balancing important?

5) Name a few assumptions that must be made to properly balance a cross section.

APPENDIX

SWAN CREEK TERM PROJECT

The Swan Creek field is located in Tennessee within the Valley and Ridge province. The northeast-trending field (fig. A–1) is situated in the footfall of the Clinchport thrust and the reservoir rocks are the Middle Ordovician Trenton and Black River groups. There are two prominent bentonites in the section that approximately mark the boundary between the two groups. This field is unusual because productive fields are typically located in the hanging wall. There are hydrocarbons (mainly gas) found in varying intervals within this field, but the goal of this project is to assess the overall structure of the footwall reservoir rocks and estimate the hypothetical oil reserves of an arbitrary interval.

Identify the stratigraphy in the 18 wells and correlate the wells with the five picks in the type log (fig. A–2). Create five structure contour maps and two cross sections in the space provided. Every well has at least one small thrust fault present in the log; indicate or note the repeated sections on the log and make sure to show the location of the fault(s) in the cross sections. (You do not need to put fault breaks in any of the maps.) Then answer the questions based on your maps and interpretations.

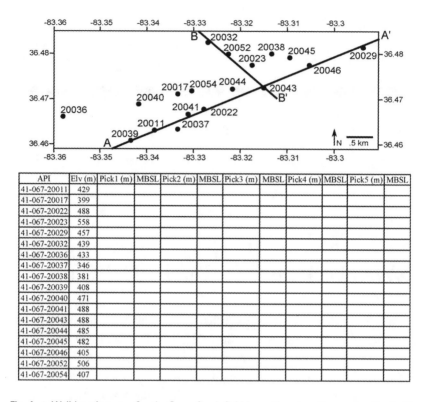

API	Elv (m)	Pick1 (m)	MBSL	Pick2 (m)	MBSL	Pick3 (m)	MBSL	Pick4 (m)	MBSL	Pick5 (m)	MBSL
41-067-20011	429										
41-067-20017	399										
41-067-20022	488										
41-067-20023	558										
41-067-20029	457										
41-067-20032	439										
41-067-20036	433										
41-067-20037	346										
41-067-20038	381										
41-067-20039	408										
41-067-20040	471										
41-067-20041	488										
41-067-20043	488										
41-067-20044	485										
41-067-20045	482										
41-067-20046	405										
41-067-20052	506										
41-067-20054	407										

Fig. A–1. Well location map for the Swan Creek field in northeastern Tennessee. *Used with permission by Tengasco, Inc.*

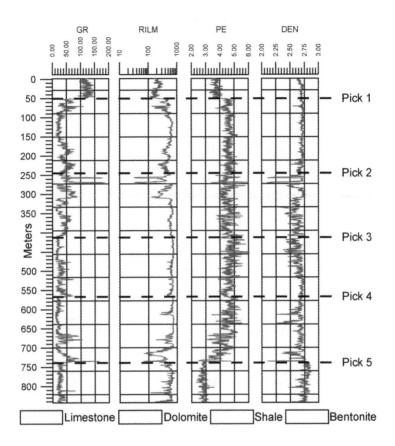

Fig. A–2. A type log through the reservoir rocks in the Swan Creek field with five picks.
Used with permission by Tengasco, Inc.

EXERCISE

Identify the stratigraphy in the wells and correlate the wells with the five picks in the type log (fig. A–2). Create five structure contour maps (figs. A–3 and A–4) and two cross sections (fig. A–5) in the space provided. Every well (figs. A–6 through A–23) has at least one small thrust fault present in the log; indicate or note the repeated sections on the log and make sure to show the location of the fault(s) in the cross sections. (You do not need to put fault breaks in any of the maps.) Then answer the questions based on your maps and interpretations.

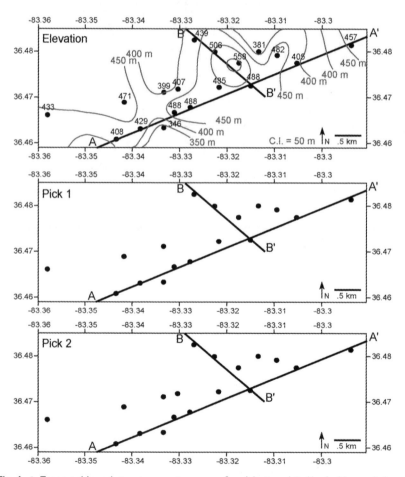

Fig. A–3. Topographic and structure contour maps for picks 1 and 2. *Used with permission by Tengasco, Inc.*

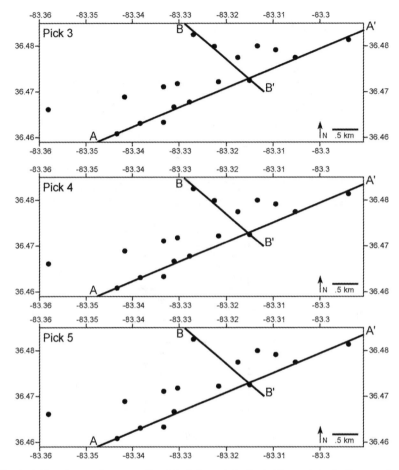

Fig. A–4. Structure contour maps for picks 1 through 3. *Used with permission by Tengasco, Inc.*

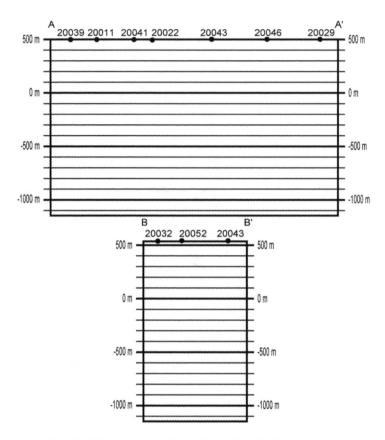

Fig. A–5. A-A' and B-B' cross sections. *Used with permission by Tengasco, Inc.*

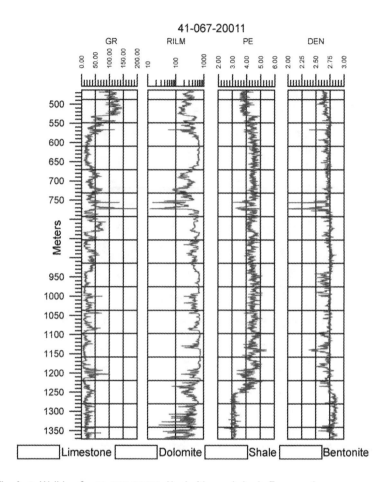

Fig. A–6. Well log for 41–067–20011. *Used with permission by Tengasco, Inc.*

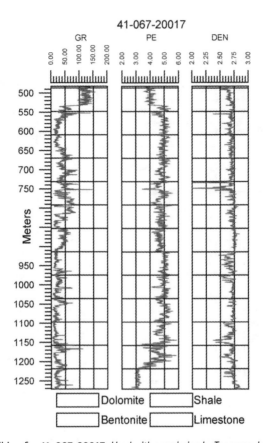

Fig. A–7. Well log for 41–067–20017. *Used with permission by Tengasco, Inc.*

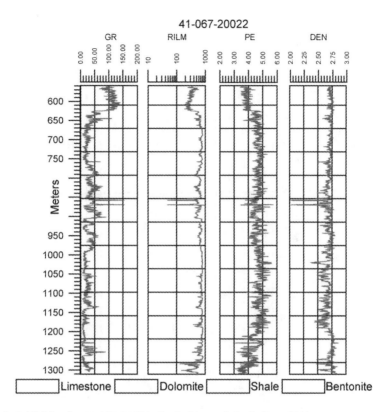

Fig. A–8. Well log for 41-067-20022. *Used with permission by Tengasco, Inc.*

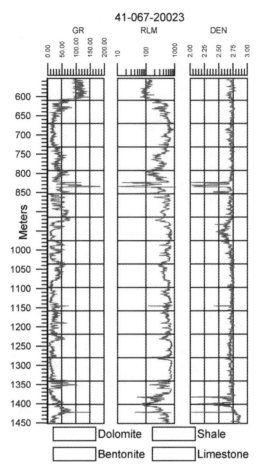

Fig. A–9. Well log for 41–067–20023. *Used with permission by Tengasco, Inc.*

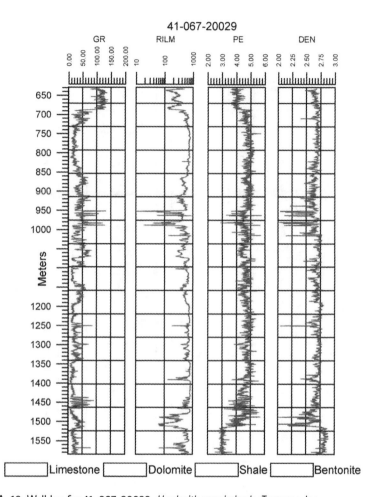

Fig. A–10. Well log for 41–067–20029. *Used with permission by Tengasco, Inc.*

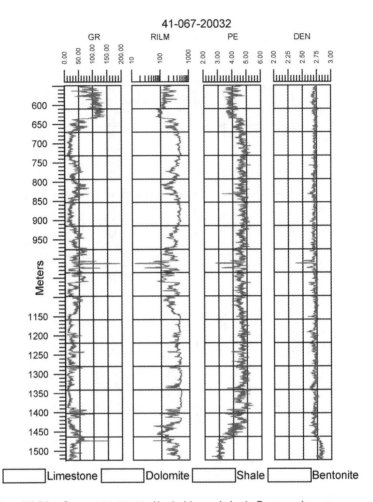

Fig. A–11. Well log for 41–067–20032. *Used with permission by Tengasco, Inc.*

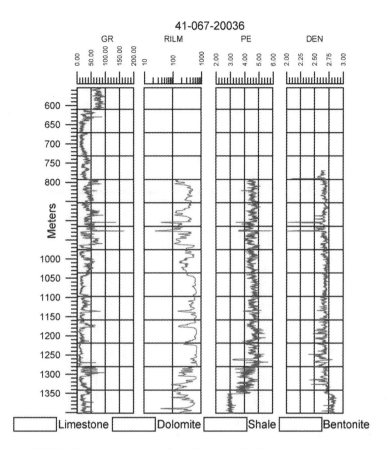

Fig. A–12. Well log for 41–067–20036. *Used with permission by Tengasco, Inc.*

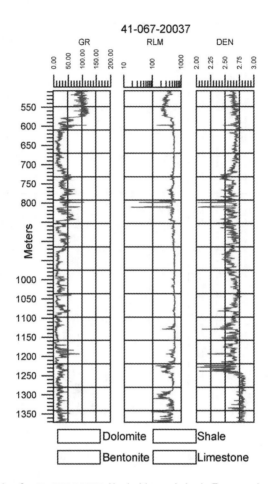

Fig. A–13. Well log for 41-067-20037. *Used with permission by Tengasco, Inc.*

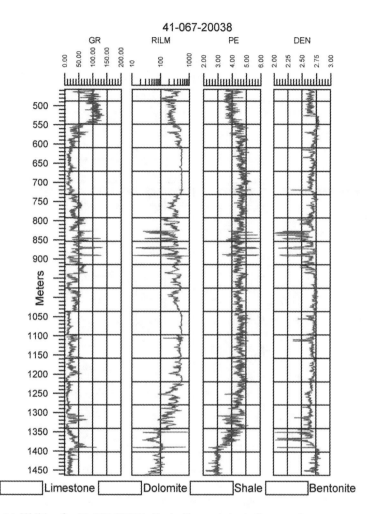

Fig. A–14. Well log for 41–067–20038. *Used with permission by Tengasco, Inc.*

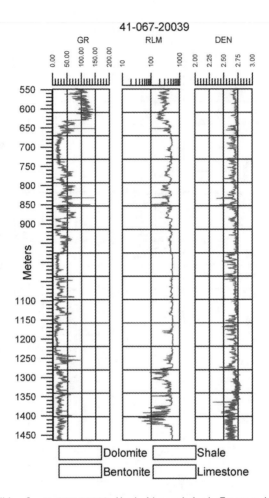

Fig. A–15. Well log for 41-067-20039. *Used with permission by Tengasco, Inc.*

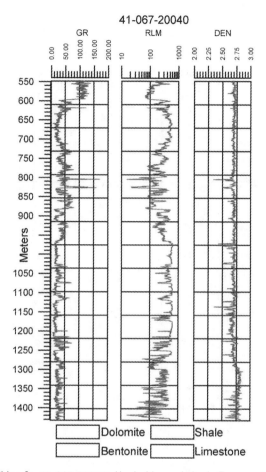

Fig. A–16. Well log for 41–067–20040. *Used with permission by Tengasco, Inc.*

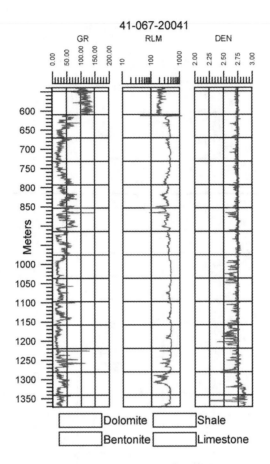

Fig. A–17. Well log for 41-067-20041. *Used with permission by Tengasco, Inc.*

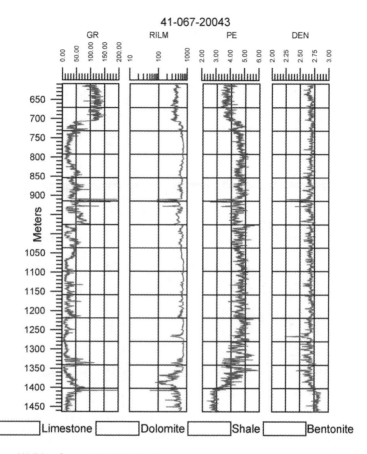

Fig. A–18. Well log for 41–067–20043. *Used with permission by Tengasco, Inc.*

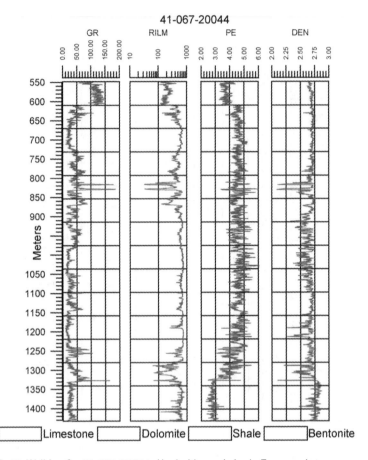

Fig. A–19. Well log for 41–067–20044. *Used with permission by Tengasco, Inc.*

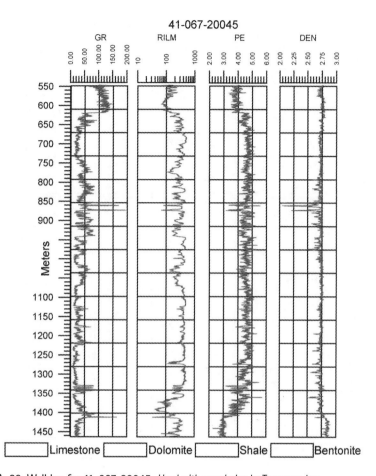

Fig. A–20. Well log for 41–067–20045. *Used with permission by Tengasco, Inc.*

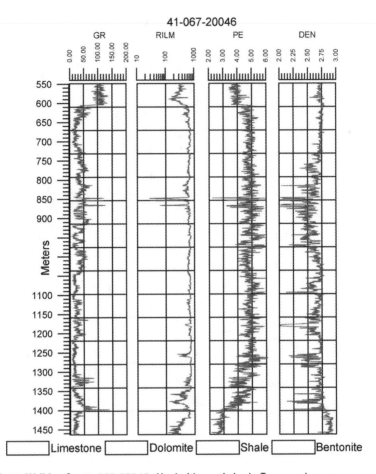

Fig. A–21. Well log for 41-067-20046. *Used with permission by Tengasco, Inc.*

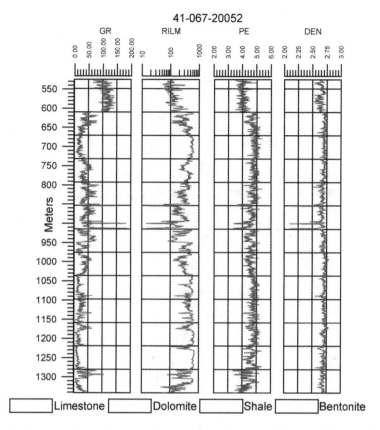

Fig. A–22. Well log for 41-067-20052. *Used with permission by Tengasco, Inc.*

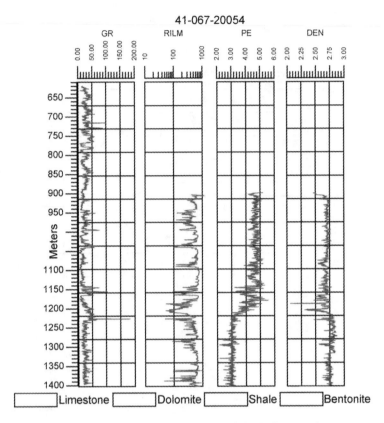

Fig. A–23. Well log for 41–067–20054. *Used with permission by Tengasco, Inc.*

QUESTIONS

1) Is this field a stratigraphic or structural play?

2) Where are most of the faults located?

3) Estimate (in m²) the aerial surface area above -400 m from the pick 2 structure contour map.

4) Assume that pick 2 represents the top of a 1.5-m thick (TVT) reservoir that is 74% water-saturated, and oil/water contact is located at -400 m. What is the volume of rock filled with oil in m³?

5) If the reservoir has on average 9% porosity, and the field has a 15% recovery rate and formation volume factor of 1, how many bbl of oil would be in this reservoir (1 m³ = 6.29 bbl of oil)?

GLOSSARY

aggradation. Deposition of sediment in a basin causing a rise in base level.

API number. Fourteen-number set that makes up a unique well identification number.

alluvial. Sediment deposited by flowing water or ice.

anchored cross section. Profile that intersect one or more wells and, therefore, has some degree of direct well control.

aquifer. Subsurface interval capable of storing or transporting water.

Archie's equation. Equation used to calculate water saturation.

area balance. Method of testing the validity of a deformed cross section by checking if areas were preserved.

atomic weight ratio. Ratio of atomic weights of the elements that make up a mineral or rock.

behind pipe. Unproduced hydrocarbon resources that remain in the reservoir.

bell pattern. Log signature that gradually increases downward and makes a bell shape.

bentonite. Type of rock or sediment composed of volcanic ash that is commonly composed of montmorillonite. Typically has a very high gamma ray signature.

biomicrite. Type of limestone consisting of minor amounts of fossil debris within a fine-grained carbonate mud matrix.

biozone. Units or intervals defined by fossil assemblages.

blanking file. An electronic section line used to calculate or return the value corresponding to each point where the section line intersects contour data.

blind structure. Structure not exposed at the surface.

borehole. Cylindrical hole drilled into the ground using a drilling apparatus.

bounded cross section. Profile that utilizes nearby data to semi-constrain the position of contacts and features.

bow pattern. Log signature that gradually decreases, then increases.

breakthrough. Time at which an injected material reaches an extraction or producing well.

bulk density. Total mass or weight of a material divided by its volume.

bull's-eye pattern. Multiple concentric isolines that make a circular map pattern.

caliper log. Log that measures the diameter of the borehole.

cased well. Well with a steel liner cemented to the sides of the borehole.

cementation. Diagenetic process by which sediment grains are bound together.

compensated neutron log. Log that records the number of hydrogen atoms in a formation related to the fluid type in the formation and formation porosity.

conductivity. Ability of a material to transmit an electrical current.

cone of depression. Conically-shaped area around a producing well where the water table or potentiometric surface is drawn down.

confined aquifer. An aquifer bound above and below by impermeable layers.

contour interval. Difference in elevation between adjacent contours.

contour line. Isoline connecting points of equal elevation.

contour slicing. A manual or electronic method for creating a profile for a given section line.

core. Cylindrical section of rock retrieved from a drilling apparatus.

cutoff. Value at which a commodity costs more to produce than the market value of the commodity.

cutting. Chips or pieces of rock retrieved from a drilling apparatus.

cylindrical pattern. Log signature that begins with a sharp kick, maintains a flat signature, then sharply returns to a baseline.

decline curve. Graph of decreasing hydrocarbon production versus time.

décollement. A detachment surface or fault that commonly prorogates parallel to bedding.

density log. Log that records the formation density in g/cm^3.

deviated well. Non-vertical borehole.

dewatering. Process of removing water from sediment via compaction or shacking.

diachronous. Rock unit or formation that has varying ages in different areas.

dipmeter log. Log from which dip magnitude and direction can be determined.

driller's log. General description of cuttings recorded as a well is drilled.

drilling floor. Base level of a drilling rig where the drillers stand.

drilling mud. Mixture of clays, barite, and water circulated around the borehole to facilitate drilling, retrieve cuttings, and prevent blowouts.

economical limit. Value at which a commodity costs more to produce than the market value of the commodity.

eolian. Processes and features related to wind activity (also referred to as aeolian).

facies. Rock type with specific characteristics that reflect a particular process or depositional environment.

fault break. Line that indicates a thrust or normal fault where data are contoured independently on either side of the break.

fluvial. Processes and features related to water or river movement.

foliation. Alignment of minerals into parallel bands or layers.

footwall. The rock underlying an inclined fault plane.

formation micro-scanner. A 360° synthetic representation or view of the borehole wall using small resistivity variations.

formation volume factor. Variable that takes into account the amount of volume loss due to dissolved gas coming out of solution during extraction.

formation water resistivity. Resistivity of the unaltered water in a formation.

funnel pattern. Log signature that gradually decreases downward and makes a funnel shape.

gamma ray log. Log that records the natural radioactivity of a formation in API (American Petroleum Institute) units.

gas effect. Crossover of neutron- and density-derived porosities due to the presence of gas in a given interval.

geothermal gradient. Ratio of temperature versus depth in the earth. Temperature increases with increasing depth.

growth faults. Fault that has contemporaneously deposited sediment overlying the structure.

hanging wall. The rock overlying an inclined fault plane.

hinge. Area near the axis of a fold.

homogenous. Unit with a uniform composition or signature.

host. Formation or unit capable of containing ore.

hydraulic head. Measurement of water pressure at a particular point in the subsurface.

in place. Total amount of a natural material in a given area.

indurated. Rock or sediment that has become hardened or lithified by pressure, heat, or cementation.

interfinger. Adjacent facies or strata that laterally pinch out or migrate back and forth.

intrasparite. Type of limestone consisting of minor amounts of intraclasts within coarse-grained sparry-calcite matrix.

isochron. Line of equal time.

isochore. Line of equal true vertical thickness.

isoline. Line of equal value.

isopach map. Map of true stratigraphic thicknesses.

isotherm. Line of equal temperature.

karst. Topography or feature formed by the dissolution of carbonate rock.

Kelly bushing. Part of a drill rig that connects to the upper end of the drill string.

kick. Prominent signature change from a baseline in a geophysical log, or the act of creating a new borehole from a preexisting borehole.

limb. Area adjacent to the hinge of a fold.

line balance. Method of testing the validity of a deformed cross section by checking if fault and stratigraphic contacts were preserved.

lithofacies. Rock type with specific characteristics that reflect a particular process or depositional environment.

marker bed. Distinctive stratigraphic unit or interval that can be traced long distances.

master borehole. Main borehole in which other deviated wells are started.

measured log thickness. The calculated thickness of an interval measured in a well log without respect to well or strata dip.

montmorillonite. Type of clay that shrinks when dehydrated and swells when hydrated.

net pay. Thickness of a potential reservoir capable of producing water or hydrocarbons.

neutron log. Log measures a formation's porosity based on the quantity of hydrogen present in the formation.

non-unique. Multiple solutions are possible for a given problem or dataset.

ore grade. Concentration of a valuable material within a given volume.

overprint. When a rock or area has undergone multiple deformational or thermal events that may remove evidence of earlier events.

palinspastic cross section. Profile that has been restored to its original, undeformed state by some combination of putting together or pulling apart fault blocks and strata.

pay zone. Interval of rock in which economical amounts of hydrocarbons are found.

permeability. Ability of rock to transmit fluids through a material.

photoelectric factor log. Log that records gamma radiation transmitted from a formation after being bombarded by photons in barns per electron (barns/e-).

photon. Electromagnetic energy or particle having no mass, charge, and an indefinitely long lifetime.

pick. Location on a well log or seismic section that correlates to an inferred stratigraphic correlation.

piezometer. A small-gauge, cased well or impermeable tube placed in the ground to measure the level to which water rises in the well.

pinch out. Disappearance of a bed or unit over a horizontal distance.

pinnacle reef. An isolated reef or coral that grows vertically.

pinned cross section. Profile that intersects one or more wells.

planimeter. Device used to measure areas.

plugged and abandoned. Filling or cementing a well and removing the drilling or production equipment.

plume. Body of contaminated groundwater.

porosity. Amount of void space within material that may be filled by fluids or gas.

porosity-foot. Total amount of void space in a given interval.

porosity log. Log that records the porosity of a given interval. There are multiple types of porosity logs such as a neutron, density, and sonic logs.

potential field. Field that obeys a differential equation known as Laplace's equation, such as gravity and magnetic fields.

potentiometric surface. Surface to which water in an aquifer can rise by hydrostatic pressure.

projected cross section. Profile with no direct well control.

provenance. Geographical area and environment from which material originated.

recumbent fold. A fold with a near-horizontal axial surface.

regression. A lowering of sea level indicated by a coarsening upward succession of strata.

reserve. Amount of profitable material that can be removed from a given location.

reservoir. Subsurface interval capable of storing or transporting fluids or gas.

resistivity. Physical property of a material to resist the flow of an electrical current.

resistivity log. Log that records the resistivity or resistance to the flow of electricity through a formation in Ohm meters (Ωm).

resource. Total amount of a natural material in a given area.

retrodeformed cross section. Profile that has been restored to its original, undeformed state by some combination of putting together or pulling apart fault blocks and strata.

rig. General term for any drilling platform.

salt dome. Mass of salt (also called a diaper) that has risen through overlying formations and caused overlying units to bulge upward.

scintillation counter. Instrument that detects and measures gamma radiation by counting the light flashes induced by the radiation.

seal. Barrier to the upward movement of hydrocarbons, resulting in the accumulation of hydrocarbons in the reservoir below.

Setchell's equation. Equation used to calculate the true vertical thickness of an interval, regardless of borehole direction or deviation, and the dip of the interval.

seismic reflection. Type of data gathered by the return of seismic energy, used to create a model of the subsurface.

shut-in. A producing well that has been stopped for some reason.

sidetrack. Deviated well drilled off the main borehole.

sole horizon. Flat plane at the base of the thrust fault.

sonde. Cylindrical string of instruments run into a borehole to generate a well log.

sonic log. Log that records the speed of sound transmitted through a formation in microseconds per foot (μs/ft).

source. Formation or location of hydrocarbon-generating rocks or material.

spill point. Lowest point in a hydrocarbon reservoir that can retain hydrocarbons.

spontaneous potential log. Log that records in millivolts (mV) the electrical current that arises due to salinity differences between a saltwater-based drilling mud and the fluid in a formation.

spud. To break ground at the site of a new well with a drilling rig.

step time. Amount of time for the fluid to travel a certain distance (also called travel time).

storm wave base. Depth at which storm waves do not affect marine sediment.

structural high. Term for features at a higher position than the surrounding features.

structural low. Term for features at a lower position than the surrounding features.

structure contour map. Map used to portray the three-dimensional surface in two-dimensions.

sweep. To push hydrocarbons toward producing wells using gas or water injected in an adjacent well.

synthetic cross section. Profile with no direct data to constrain the position of contacts and features.

tadpole plot. Plot of dipmeter data where the dip magnitude is plotted as a circle on the horizontal axis and the line segment coming off the circle points in the direction of dip.

temperature log. Log that records the temperature within the borehole.

three-point problem. Technique to calculate the strike and dip of a surface from three data points.

time line. Line that represents a given time.

tortuosity. The amount of curvature in a material related to the overall distance a fluid must travel to get through a given interval.

tracks. Sections on a well log where particular logs are plotted.

trap. Barrier to the upward movement of hydrocarbons, resulting in the accumulation of hydrocarbons in the reservoir below.

transgression. A rising of sea level indicated by a fining upward succession of strata.

travel time. Amount of time for the fluid to travel a certain distance (also called step time).

trend surface map. Map of a fitted plane to a corresponding dataset.

trend surface residual anomaly map. Map of the residual values calculated by subtracting a trend surface from the original data.

true stratigraphic thickness. Thickness (perpendicular to bedding) of an interval.

true vertical depth thickness. Vertical thickness of an interval along a borehole from the entry point to the exit point.

true vertical thickness. Vertical thickness of an interval.

turbidite. Sediments deposited by submarine turbidity currents.

uncased well. Well without a liner.

unconfined aquifer. An aquifer that has no confining layers above it.

uninvaded. Section of a formation not disrupted by drilling.

vertical exaggeration. Dimensionless number that compares the horizontal and vertical scales of a cross section.

volume balance. Method of testing the validity of a deformed cross section by checking if volumes were preserved.

wash out. Channel-like cavity formed by strata surrounding the borehole moving or flowing into the borehole.

Walther's law. Coeval facies deposited laterally and facies stacking order reflecting changes in depositional environments through time.

water saturation. Amount or percentage of water in an interval.

water table. Surface that separates the vadose, or unsaturated zone, from the saturated zone.

well log. Detailed record of the geophysical and physical properties in and around the borehole obtained during drilling.

well ticket. Paper that contains a synopsis of the important drilling information.

well-to-well cross section. Profile that directly correlate wells.

wildcat. Exploratory well drilled in a region with little or no data.

LIST OF ABBREVIATIONS

a – Tortuosity factor

API – American Petroleum Institute

AWR – Mineral weight ratio

barns/e- – barns per electrion

bbl – barrels

Bcf – Billion cubic feet

Cal – Caliper

cf – Cubic foot

C.I. – Contour interval

CNC – Compensated neutron correction

CNL – Compensated neutron log

d – Day

Den – Density

D.F. – Drilling floor

DPHI – Density porosity

e- – Electron

FBSL – Feet below sea level

ft – Feet

FMS – Formation micro-scanner

FVF – Formation volume factor

g – Gram

G.L. – Ground level

GR – Gamma ray

in – Inch

K.B. – Kelly bushing

m – Meter

m – Cementation exponent

MBSL – Meters below sea level

Mcf – Thousand cubic feet

md – Millidarcy

MLT – Measured log thickness

MMbbl – Million barrels

MMcf – Million cubic feet

mV – Millivolt

n – Saturation exponent

OG – Ore grade

PE – Photoelectric factor

RILD – Deep induction resistivity

RILM – Medium induction resistivity

RR – Recovery rate

R_0 – Water saturated formation resistivity

R_t – Formation resistivity

R_w – Water resistivity

SL – Sea level

SF – Sea floor

SP – Spontaneous potential

S_o – Oil saturation

S_w – Water saturation

Temp – Temperature

TSRA – Trend surface residual anomaly

TST – True stratigraphic thickness

TVD – True vertical depth

TVDT – True vertical depth thickness

TVT – True vertical thickness

VE – Vertical exaggeration

w – Well

Yr – Year

θ_a – Apparent horizon dip

θ_b – Borehole dip

θ_t – Azimuth horizon

θ_w – Azimuth borehole direction

° – Degree

°F – Degree Fahrenheit

µs – Microsecond

Ω – Ohm

Φ – Porosity

BIBLIOGRAPHY

Asquith, G., and D. Krygowski. 2004. *Basic well log analysis*. American Association of Petroleum Geologists.

Bohling, G., and J. Doveton. *The Oz machine*. Online tutorial on geological interpretation of wireline logs. Kansas Geological Survey. http://www.kgs.ku.edu/PRS/ReadRocks/OzIntro. html.

Boggs, S., Jr. 2005. *Principles of sedimentology and stratigraphy*. New York: Prentice Hall.

Davis, G. H., and S. J. Reynolds. 1996. *Structural geology of rocks and regions*. New York: Wiley.

Davis, J. C. 2002. *Statistics and data analysis in geology*. New York: John Wiley & Sons.

Hearst, J. R., P. H. Nelson, and F. L. Paillet. 2000. *Well logging for physical properties: A handbook for geophysicists, geologists, and engineers*. New York: Wiley.

Hyne, N. J. 2001. *Nontechnical guide to petroleum geology, exploration, drilling, and production*. Tulsa: PennWell.

Johnson, D. E., and K.E. Pile. 2002. *Well logging in nontechnical language*. Tulsa: Pennwell.

Keys, W. S. 1990. Borehole geophysics applied to ground-water investigations. *U.S. Geological Survey.* Techniques of water resources investigations report 02-E2.

Krajewski, S. A., and B. L. Gibbs. 1994. Computer contouring generates artifacts. *Geotimes.* 39(4):15–19.

Serra, O. 1984. *Fundamentals of well log interpretation: The acquisition of logging data.* London: Elsevier Science Ltd.

Tearpock, D. J., and R. E. Bischke. 2003. *Applied subsurface geological mapping with structural methods.* New York: Prentice Hall.

Twiss, R. J., and E. M. Moores. 2006. *Structural geology.* New York: W. H. Freeman.

INDEX

A

abandoned wells, 149, 216
abbreviations, 221–224
aggradation, 23, 211
alluvial, 93, 211
American Petroleum Institute (API), 5
anchored cross sections, 170, 172, 211
angular unconformity, 108
anomalous resistivity spikes, 139
anticlinal trap, doubly plunging, 151
API number, 211
    county code in, 6
    event code in, 7
    sidetrack code in, 7
    as well ID number, 5
    well number as part of, 6–7
aquifers, 41, 137–138, 211–212, 220
Archie's equation, 153, 211
area balance, 175, 211
area of closure, 154
asymmetric patterns, 23
atomic weight ratio (AWR), 163, 211

B

balanced cross section, 175–177
bedding planes, 45
behind pipe, 154, 157, 211
bell patterns, 23, 211
bentonite, 41, 211
biomicrite, 93, 211
biozones, 91, 211
black box approach, 36
blanking file, 172, 212
blind structures, 119, 212
boreholes, 208
    caliper log recording, 22
    geophysical instruments placed
        in, 1
    sea level changes interpreted
        with, 91–92
bounded cross sections, 170, 212
bow patterns, 23, 212
breakthrough point, 140, 142, 212
brines, 139
bulk density, 21, 212
bulk lithology, 20
bull's eye patterns, 46, 212

C

caliper logs, 2, 22, 212
cased well, 138, 212
cementation, 212
cementation values, 152
codes
    county, 6
    event/sidetrack, 7
    state, 5

compensated neutron logs (CNL), 21, 212

composite logs, 45

compression velocities, 22

computers
    contour map from, 36
    trend surface modeled by, 107

conductivity, 20, 212

cone of depression, 140, 212

confined aquifers, 138, 212

consolidated rocks, 22

contaminants, 138–139

contour intervals, 34, 212
    data variance and, 36
    scale-dependent maps and, 58–59

contour line, 208

contour maps
    computers used for, 36
    map data representation in, 34
    three-stacked structure in, 59

contour slicing, 172, 174, 212

conversion chart, 77

core, 93, 140, 161, 213

correlations
    stratigraphic, 18
    subsurface, 10

county code, in API number, 6

cross sections, 186
    anchored, 170, 172, 211
    balanced/unbalanced, 175–177
    categories of, 170
    detailed, 173
    different types of, 171
    of fold, 74
    of injection wells, 141
    pinned, 172, 216
    retrodeformed/palinspastic,
        175–176, 216–217

of salt dome, 75

of structural high, 42

of subsurface, 169

validity of, 177

vertical exaggeration in, 174–175

vertical/horizontal comparison
    in, 174

of water table, 139

well-to-well, 172, 220

cutoff, 75, 213

cutoff line, 157

cutting, 93, 213

cylindrical patterns, 23, 213

D

data
    contour interval variance with, 36
    state cataloging differences of, 5

decline curves, 155–156, 213

décollement, 38, 213

density logs, 2, 21–22, 150, 213

density porosity, 76

depositional environments, 94

detailed cross section, 173

deviated wells, 9, 213

dewatering features, 46, 213

diachronous, 91, 213

digital files, 1

dip measurements, 22

dipmeter logs, 22, 213

driller's logs, 137, 213

drilling
    logging data while, 1
    mud, 213
    well tickets information on, 2

drilling floor (DF), 2, 213

E

economic limit, 157, 213
electric logs, 17
electrical currents record, 20
electrical flow record, 20
elevation maps, subsurface, 58
eolian, 93, 213
event code, in API number, 7
exploration wells, 112–116

F

facies, 213
    generating map of, 93
    lateral, 46
    pay zone and, 93
    similar characteristics in, 91
fault break, 34, 35, 174, 213
fault-controlled ore deposit, 162
faults, 43–45
feet below sea level (FBSL), 57
first-order
    trend surface map, 122
    TSRA, 122
flat patterns, 23
flowchart, of TSRA, 123
fluvial, 93, 214
folds, 43–44, 74, 217
foliation, 45, 214
footwall, 177, 214
formation fluid type, 149–150, 150
formation micro-scanner logs (FMS), 45–46, 214
formation volume factor (FVF), 155, 214
formation water resistivity, 152, 214

formations porosity measurements, 21–22
funnel patterns, 23, 214

G

gamma radiation transmission, 20
gamma ray logs (GR), 2, 150, 214
    multiple picks in, 43
    natural radioactivity recorded in, 18
gas effect, 21, 149, 214
gas/oil contacts, 150–151
geologic maps, 174
geophysical instruments, 1
geophysical logs
    structural features identified in, 43–44
    of test wells, 161
    thrust fault identified in, 121
geophysical properties
    of subsurface strata, 1
    in well log, 2
geophysical responses, 150
geothermal gradient, 214
ground level (GL), 2
groundwater flow
    cone of depression from, 140
    modeling of, 137
    travels of, 139
growth faults, 177, 210
growth strata, 177

H

hanging wall, 177, 214
heterogeneity, 140

hinge, 44, 176, 214
homogeneous, 41, 214
horizontal comparisons, 174
host, 162, 214
hydraulic head, 214
    groundwater travels and, 139
    as water pressure measurement, 138
hydrocarbon plays, 161–162
hydrocarbon reserves
    calculating, 154–157
    estimating potential, 153
    identifying, 149
    porosity measurements
        estimating, 140
hydrocarbon-filled interval, 151
hydrologic maps, 137–139

I–J

ideal maker beds, 41
in place, 140, 214
indurated, 22, 214
injection wells, 140–142
interfinger, 93, 211
interval
    contour, 34, 58–59, 212
    contour variance of, 36
    hydrocarbon-filled, 151
    net pay/pay zone and, 75
    thickness, 8
intrasparite, 93, 215
irregular patterns, 23, 46
isochore, 215
isochore maps, 74
isochron, 34, 215
isolines, 34, 215
isopach maps, 33, 215
    of fold, 74

    irregular patterns in, 46
    stratigraphic interpretations from, 45–46
    thickness map misinterpreted as, 73
isotherm, 215

K

karst, 74, 215
Kelly bushing (KB), 2, 215
kick, 5, 23, 215

L

lateral facies, 46
limb, 73, 215
limestone, densities of, 76–77
line balance, 175, 215
linear grids, 4
linear thickening trends, 74
lithofacies, 91, 215
lithologies, 20
    in exploration wells, 112–116
    in well logs, 25–30, 39, 47–54,
        60–71, 78–89, 95–105,
        124–135, 158
log signature patterns, 23
logarithmic grids, 4
logging data, 1

M

map data representation
    in contour map, 34
    of facies, 93
mappable horizons, 41
mapping skills, subsurface, 109

maps. *See also* Structure contour
    maps; Trend surface maps
    contour, 34, 36, 59
    facies, 93
    first-order trend surface, 122
    generating facies, 93
    geologic, 174
    hydrologic, 137–139
    isochore, 74
    isopach, 33, 45–46, 73–74, 215
    net pay, 75
    scale-dependent, 58–59
    schematic, 141
    stacked structure contour, 120
    stacked TSRA, 120
    subsurface, 33, 43–44, 161
    subsurface elevation, 58
    temperature, 35
    thickness, 73, 75
    three-stacked structure in, 59
    trend surface residual anomaly,
        33, 120–123, 219
    types of, 33
marker bed, 33, 41, 57, 215
marker horizons, 173
marker unit, 171
master borehole, 215
measured log thickness (MLT), 8–9,
    215
measurements
    conversion chart of, 77
    in density log, 21–22
    dip, 22
    formations porosity, 21–22
    hydrocarbon reserves estimated
        in, 140
    porosity, 140, 216
    of sedimentary rocks, 94
    strike, 22

    subsurface strata thickness, 8–11
    void space in, 161
    water pressure, 138
meters below sea level (MBSL), 57
mineral deposits, 161–162
mineral resource system, 162
mineralization, 162
mining deposit reserve, 163–164
mining industry, 161
montmorillonite, 22, 215
mud, drilling, 213

N

natural radioactivity, 18
net pay, 211
    in given interval, 75
    maps, 75
    values calculated of, 76–77
    well logs showing, 76
neutron logs, 2, 21–22, 150, 216
non-unique data, 58, 216

O

oil resources, calculating, 154–155
oil saturation, 153
oil/water contacts, 150–151
ore grade, 161, 216
ore grades, 165–166
outcrop, subsurface, 169
overprint, 58, 216

P–Q

palinspastic cross section, 175–176, 216
paper logs, 1

pay zone, 216
    facies map for, 93
    hydrocarbons identified in, 149
    interval/reservoir potential for, 75
permeability, 216
    recoverable reserves estimated
        by, 140
    of sedimentary rocks, 94
photoelectric logs (PE), 2, 20, 150, 216
photon, 20, 216
physical properties, in well log, 2
picks, 216
    in GR logs, 43
    in structure contour map, 184–185
    Swan Creek project with, 183
piezometer, 138, 216
pinch out, 93, 216
pinnacle reef, 74, 216
pinned cross section, 172, 216
planimeter, 154–155, 216
plugged back wells, 8
plugged wells, 149, 216
plumes, 138–139, 216
porosity logs, 2, 76, 217
porosity measurements, 216
    conversion chart of, 77
    in density log, 21–22
    hydrocarbon reserves estimated
        in, 140
    of sedimentary rocks, 94
    void space of, 161
porosity-foot, 77, 217
porous reservoir horizon, 152
potential field, 33, 109, 217
potential reserves, 164
potential resources, 163–164
potentiometric surface, 138, 217
production data, 156
productive wells, 149

projected cross sections, 170, 217
provenance, 107, 217

R

radioactive logs, 17
recoverable reserves
    calculating, 154–157
    permeability estimating, 140
    spill points estimations with,
        154–155
recumbent fold, 44, 217
regional dip
    three-point problem representing,
        109–111
    trend surface map identifying,
        45–46
regional stratigraphy
    changes in, 107
    understanding, 41
    from well data, 2
regional subsurface framework, 41
regression, 91–93, 217
regression matrix, 121
regressional changes, 92–93
reserve, 213
    calculating mining deposit,
        163–164
    calculating recoverable, 154–157
    injection well recoverable, 140
    water, 75
reserve volume, 164, 217
reservoirs, 137, 217
    analysis, 152
    heterogeneity, 140
    pay zone potential in, 75
resistivity, 2, 20, 217
resistivity logs, 217
    electrical flow recorded in, 20

water table identified in, 138
resource, 217
    calculating mining, 163–164
    calculating oil, 154–155
retrodeformed cross section, 175–176, 213
rig, 1, 217

S

salt dome, 74, 75, 217
sandstone, densities of, 76–77
Savannah, GA, 43
scale-dependent maps, 58–59
schematic maps, 141
scintillation counter, 18, 217
sea level
    boreholes interpreting changes in, 91–92
    transgressional/regressional changes in, 92–93
seal, 154, 218
sedimentary rocks, 94
seismic contours, 154–155
seismic reflection, 2, 33, 57, 218
Setchell's equation, 8, 218
shut-in, 154, 157, 218
sidetrack, 7, 218
signature patterns, of well logs, 19, 24
small-scale fault, 58–59
sole horizons, 119, 218
sonde, 1, 218
sonic logs, 2, 22, 218
sound, 22
source, 46, 74, 91–93, 218
spill points, 154–155, 218
spontaneous potential logs (SP), 2, 150, 218
    electrical currents recorded in, 20

water table identified in, 138
spudded, 5, 218
stacked structure contour maps, 120
stacked TSRA maps, 120
state(s)
    codes, 5
    well data cataloging differences of, 5
    wells in, 6
step time, 140, 218
storm wave base, 218
stratigraphic contacts, 41
stratigraphic correlations, 18
stratigraphic interpretations, 45–46
stratigraphic thickness, 73
stratigraphic traps, 94
strike measurements, 22
structural features, 43–44
structural high, 42, 218
structural interpretations, 43–45
structural logs, 17
structural low, 44, 218
structure contour maps, 33, 219
    of fold, 74
    hydrocarbon-filled interval shown in, 151
    multiple folds in, 43–44
    picks in, 184–185
    stacked, 120
    subsurface investigated with, 57
    of thrust fault, 35
    wireline mesh representation of, 58
subsurface
    correlations, 10
    cross section of, 169
    elevation map, 58
    mapping skills, 109
    outcrop, 169

structure contour map
  investigating, 57
subsurface maps, 33
  mining industry using, 161
  structural features identified in,
    43–44
subsurface strata
  geophysical properties of, 1
  map types of, 33
  thickness measurements of, 8–11
  understanding regional, 41
Swan Creek project, 185–209
  five picks log of, 187
  well location map of, 186
sweep, 219
sweep area, 142
symmetric patterns, 23
syndeformational sedimentation, 177
synthetic cross sections, 170, 219

T

tadpole plot, 22, 219
temperature logs, 2, 23, 138, 219
temperature maps, 35
test wells, geophysical logs of, 161
thickness maps
  net pay maps as, 75
  stratigraphic thickness
    represented in, 73
thickness measurements, 8–11
thickness values, 11
3-D seismic reflection data, 119
3-D seismic surveys, 161
three-point problem, 219
  solving, 109–111
  trend surface map generated from,
    109–111
three-stacked structure, 59

thrust fault
  geophysical log identifying, 121
  structure contour map with, 35
  vertical exaggeration and,
    174–175
time lines, 91, 93, 219
topography
  maps of, 184
  trends in, 58
tortuosity, 152, 219
tracks, 215
  linear/logarithmic grid numbers
    of, 4
  of well log, 2
transgressional changes, 92–93
transgressions, 91–93, 219
trap, 219
travel time, 139–140, 219
trend surface maps
  angular unconformity in, 108
  computers modeling, 107
  first-order, 122
  regional dip identified in, 45–46
  regional stratigraphic changes
    understood through, 107
  regression matrix used in, 121
  three-point problem generating,
    109–111
trend surface residual anomaly maps
  (TSRA), 33, 219
  first-order, 122
  flowchart of, 123
  making, 121
  stacked, 120
  3-D seismic reflection data and,
    119
true stratigraphic thickness (TST), 219
  calculating, 10
  interval thickness as, 8

true vertical thickness v., 11
true vertical depth thickness, 219
true Vertical depth thickness (TVDT), 8, 10
true vertical thickness (TVT)
　calculating, 8–10, 219
　true stratigraphic thickness v., 11
turbidite, 46, 220
turbiditic shales, 46

U

unbalanced cross section, 175–177
uncased holes, 17
uncased well, 138, 220
unconfined aquifers, 138, 220
uninvaded, 20, 220

V

vertical comparisons, 174
vertical exaggeration (VE), 174–175, 220
vertical scale, 174
vertical wells, 9
void space, 161
volume balance, 175, 176, 220

W–Z

Walther's law, 91, 220
wash out zones, 18, 220
water pressure measurement, 138
water saturation, 152–153, 220
water table, 137, 220
　groundwater travels and, 139
　resistivity logs identifying, 138
　SP log identifying, 138

well(s)
　decline curves of, 155–156
　economic limit of, 157
　ID number (API number) of, 5
　ore grades contoured in, 165–166
　plugged back, 8
　production data of, 156
　same name/different states of, 6
　sidetracks of, 7
　Swan Creek project map of, 182
　thickness values misused in, 11
　tickets of, 2, 220
　vertical/deviated, 9
well data
　regional stratigraphy from, 2
　state cataloging differences of, 5
well logs, 220. *See also* Caliper logs;
　Compensated neutron logs;
　Composite logs; Density logs;
　Dipmeter logs; Driller's logs;
　Electric logs; Formation micro-
　scanner logs; Gamma ray logs;
　Geophysical logs; Neutron logs;
　Paper logs; Photoelectric logs;
　Porosity logs; Radioactive logs;
　Resistivity logs; Sonic logs;
　Spontaneous potential logs;
　Structural logs; Temperature logs
　composite type in, 45
　contaminants identified by, 138–139
　first page of, 3
　of 41-067-20011, 191
　of 41-067-20017, 192
　of 41-067-20022, 193
　of 41-067-20023, 194
　of 41-067-20029, 195
　of 41-067-20032, 196
　of 41-067-20036, 197

of 41-067-20037, 198
of 41-067-20038, 199
of 41-067-20039, 200
of 41-067-20040, 201
of 41-067-20041, 202
of 41-067-20043, 203
of 41-067-20044, 204
of 41-067-20045, 205
of 41-067-20046, 206
of 41-067-20052, 207
of 41-067-20054, 208
gamma ray, 18
geophysical/physical properties
    in, 2
lithology identification in, 25–30,
    39, 47–54, 60–71, 78–89,
    95–105, 124–135, 158
main categories of, 17
measured log thickness in, 8–9
net pay shown in, 76
signature patterns of, 19, 24
tracks of, 2
well number, as part of API number,
    6–7
wellbore temperature log, 23
well-to-well cross section, 172, 220
wildcat, 109, 220
wireline mesh representation, 58